BEEKEEPING

FOR BEGINNERS

DR. JONATHAN STEELE

In this book, we will discuss various aspects of beekeeping, including the historical and cultural significance of beekeeping, the scientific study of bee behavior, as well as the practical application of hive management. Additionally, we will examine current research on honeybee health and address some of the challenges that modern beekeepers face.

TABLE OF CONTENTS

INTRODUCTION TO BEEKEEPING

Beekeeping is a fascinating and rewarding practice that has captivated people for centuries. Whether you are an experienced beekeeper or a beginner just starting out, there is always something new to learn about these fascinating insects and the art of beekeeping. In this book, we will take a deep dive into the world of beekeeping, exploring everything from the history of beekeeping to the science of bee behavior and the practicalities of managing a hive. One of the most compelling reasons to learn about beekeeping is the crucial role that bees play in our ecosystem. Bees are responsible for pollinating a vast number of plant species, including many of the fruits and vegetables that we rely on for sustenance. Without bees, many of these crops would fail to produce the fruits and vegetables that we need to survive. Moreover, bees are also an essential part of the biodiversity of our planet, pollinating a wide range of wildflowers and other plants that support the health and resilience of our ecosystems. In addition to their critical role as pollinators, bees also provide us with a range of valuable products, from honey and beeswax to propolis and royal jelly. Honey, in particular, has been prized for its health benefits and culinary uses for centuries, and beekeeping is a valuable source of income for many people around the world. However, beekeeping is not without its challenges. In recent years, honeybee populations have been under threat from a range of factors, including disease, habitat loss, and the use of pesticides. Managing a hive requires a significant amount of skill and knowledge, and beekeepers must be able to identify and address a range of potential issues to keep their hives healthy and productive. Beekeeping is that one significant practice that has been passed down through generations, spanning across continents and cultures. From the earliest forms of beekeeping, where bees would be kept in hollowed-out logs or bonesets, to the modern, standardized hives we see today, beekeeping has always been an artful and valuable practice. The importance of beekeeping cannot be overstated, as it is vital to our environment, our food supply, and our economy. In fact, it is estimated that bees are responsible for pollinating approximately one-third of the food we eat. This is why bees are often referred to as the "unsung heroes" of our food supply. In addition to their role as pollinators, bees also play an essential role in the preservation of biodiversity. Bees are an integral part of the ecosystem, pollinating a vast number of plant species, including many wildflowers. This helps to maintain the diversity of plant species, which, in turn, supports the health of our ecosystems. Furthermore, by supporting the growth of wildflowers and other plants, bees contribute to the overall health and resilience of the ecosystem. Beekeepers can earn income from the sale of honey, beeswax, and other bee-related products. In addition, beekeeping can support local agriculture by providing pollination services to farmers and increasing the productivity of their crops. This makes beekeeping an essential practice in many rural communities, as it can help to support local economies and provide a source of income for families.

Perhaps most importantly, beekeeping can help to protect honeybee populations, which are under threat from a variety of factors, including pesticides, disease, and habitat loss. By raising healthy bees in controlled environments, beekeepers can help to support honeybee populations and contribute to their preservation. In addition, beekeepers can monitor the health of their hives and take steps to prevent disease outbreaks, which can help to protect honeybee populations from the devastating effects of colony collapse disorder. Beekeeping is not just an ordinary practice but an art that requires a vast array of skills, knowledge, and expertise. Beekeepers need to have a thorough understanding of bee behavior, hive management, disease prevention, and honey harvesting techniques. They must be able to identify the different types of bees in their hives and understand their roles within the hive. They also need to have an excellent understanding of plant species and their blooming times to ensure that their bees have access to enough food throughout the year. The art of beekeeping has evolved over time, with new techniques and technologies emerging to improve the health and productivity of hives. From the development of new hive designs and materials to the use of drones and sensors to monitor hive health, beekeeping is a practice that is constantly evolving and improving. In addition to its practical benefits, beekeeping also has cultural and symbolic significance. Bees have been revered throughout history, with ancient civilizations considering them to be sacred beings associated with fertility, wisdom, and communication. The hardworking nature of bees has also made them a symbol of diligence and industry, inspiring people to work hard and be productive.

Throughout this book, we will explore the many facets of beekeeping, from the history and culture of beekeeping to the science of bee behavior and the practicalities of hive management. We will also delve into the latest research on honeybee health and explore some of the challenges facing beekeepers today. Whether you are a seasoned beekeeper or someone who is simply interested in learning more about these fascinating insects, this book is for you. We hope that you will find this book to be an engaging and informative resource that inspires you to deepen your knowledge of beekeeping and appreciate the vital role that bees play in our world. So join us on this journey into the fascinating world of beekeeping, and discover the beauty and wonder of these essential insects.

Chapter 1

BEGINNING YOUR JOURNEY AS A BEEKEEPER

Many new beekeepers come into this intriguing undertaking with high expectations, and this approach colors their entire mentality. The art seems, by all accounts, to be easygoing and seems to be fun, so why not give it a shot? Albeit pleasant, beekeeping can be a disheartening experience. It requires arrangement and responsibility. It requests information on the regular world. The beekeeper likewise profits by a specific measure of connection with others in the apicultural local area. Like the bees in their state cooperating to get by, no singular human can succeed alone with regards to really focusing on this social bug. Beginning with two settlements permits you to look into it.

Key Practices for Launching Beekeeping Adventure

It may seem overwhelming to the beginner, with an overabundant appeal inventory on the most proficient method to guarantee a positive outcome. You can't have a clue about every one of the traps ahead of time, yet you might find yourself disinclined to start and experiencing an instance of "investigation loss of motion," attempting to realize everything before beginning. When a moderate measure of data has been processed, nonetheless, it's best to plunge in. The accompanying tips are recommended to facilitate the way.

1. Begin with new gear of standard (Langstroth) plan and aspects. Utilized custom-made made hardware by making issues that the amateur isn't ready to perceive or deal with. 2. Try not to analyze during your most memorable little while. Learn and utilize essential strategies. Ace them. Once more, you will have a reason for correlation if you decide to try in later years. 3. Before purchasing a purported fledgling outfit, realize how each piece of hardware is utilized and be certain it is required. 4. Begin with Italian bees. They are the norm in the US and are generally regularly accessible. Those procured from an equipped maker ought to be delicate and simple. In later years, you can explore different avenues regarding different races or strains to frame examinations. 5. Begin with a bundle of bees or a core settlement (nuc) instead of a laid, completely populated one. Laying out a nuc or bundle will assist with acquiring certainty. Please, begin with two settlements to give a further premise of examination. The weaknesses of one ought to be more straightforward to recognize, and a subsequent unit can give an important asset — bees and brood to keep your beekeeping activity alive in the event of a crisis. 6. Begin from the get-go in the season, yet all the same, not too soon. Look for directions from neighborhood beekeepers and controllers about timing in your region. Recollect that all beekeeping is nearby. Regardless of whether they are fruitful, counsel about overseeing bumble bees from those in other geographic regions is loaded with risk. 7. Perceive that your states won't create an overflow of honey the primary year, particularly those created from bundle bees. The main year is a learning time for the beekeeper and a structured time for the new bumble bee states.

Beekeeping is diverse and includes significantly more than setting up a hive on a patio, visiting it several times each year, and receiving the benefits. It can require an extraordinary arrangement concerning exertion, time, and cash. The fledgling can undoubtedly be overpowered by shuffling every one of the balls expected to deal with a bumble bee settlement. The best game plan is to have an arrangement, begin little, and step by step progress in discrete augmentations, always cautiously checking your objectives and changing them as conditions warrant.

Of every single agrarian endeavor, beekeeping is one of the hardest to make due to. Most animals are under a proportion of control. Dairy cattle, ponies, and pigs can be fastened, fenced, or bound. Crops don't move, albeit this can likewise be a drawback concerning their administration. Quantifying the feed, cash, time, and exertion, a supervisor could set into most farming honeys as moderately simple against what is delivered. Bumble bees, conversely, are free-flying bugs, and a decent extent of a settlement's people might be rummaging inside a 2-mile span of the hive. The beekeeper usually needs more information on where the bees are or what they may do.

The time given to keeping bumble bees can be imperfect; likewise, with numerous exercises, you get back relatively what you put into it. The time responsibility likewise shifts with the beekeeper's objectives. The biggest time speculation will be in the learning stage as you start the art: visiting bee schools, studios, and gatherings of the neighborhood and public affiliations. Anticipate an expectation to absorb information that will be steep the primary year or somewhere in the vicinity, and keep expecting difficulties throughout your beekeeping profession. Numerous people attempt beekeeping with negligible readiness, accepting they can make a plunge and get the imperative information. More possible previously, this has turned into an inexorably less effective methodology as bumble bees have become substantially more helpless against the extraordinary irritations and sicknesses that torment them and the beekeeper.

Start With a Pair

Starting with two bee colonies rather than one would be a great idea. The justification behind this is the genuine chance that, as a fledgling, you will lose a state in the principal season. Having one for later use will give you a pad against a total debacle that could finish your beekeeping exercises. Two colonies will likewise be helpful for similar purposes. Plan on visiting the bees normally like clockwork during the dynamic season, maybe more frequently as the new season starts, and now again in the idle piece of the year. Individual assessments per hive can be concise, contingent upon the season and justification for being there. Some might most recent a little while; others, including a particular errand, could require 20 to 30 minutes for every hive at the longest. Most examinations that include opening the hive are a significant disturbance to settlement life and thus should fill a significant need. Have an objective as a primary concern before upsetting the bugs. Bee states ought never to be completely disregarded. A few undertakings can be gotten or put down voluntarily. Not beekeeping. Errands not done at the legitimate time aren't only possible later, if by any means.

Visit the bees like clockwork or so during spring and summer, consistently in light of a reason and an objective. Record Keeping Most beekeepers need to keep sufficient records. It's just basic. Complete long-range bee the executives and monetary record-keeping systems are fundamental to augment productivity and keep up with benefit in any apicultural venture. Visiting any bee yard typically uncovers settlements set apart by sticks, stones, or other promptly accessible materials on every state's cover. The particular game plan of these materials might show everything from the status of the sovereign to whether a settlement requires food. Two issues emerge with this sort of record keeping. It is short-range; when the situation with the settlement changes and the materials are improved, the past data is lost. Likewise, the System is novel to every administrator, delivering it as nontransferable and untranslatable to others. Part-time representatives or bee reviewers will probably need to learn the significance of such records. It is additionally crucial for Imprint sovereigns. It allows the beekeeper to pass judgment on the efficiency of people bought from various reproducers to decide a sovereign's age and whether supersedure has happened. The most serious issue with record keeping is choosing what data to save. The sheer volume of potential information that can be gathered could be better. The beekeeper must, subsequently, cautiously pick what is generally significant for every specific venture.

Go into beekeeping rigorously as an agreeable leisure activity. Anticipate that it should cost cash at first. It will be all outgo for some time. That is awesome if you have excess honey to sell after two or three years. You may have enough hives to move quickly into crop fertilization.

At last, if everything is working out in a good way, you could transform this side interest into a sideline business, yet wait to put together your future concerning it. Hold on until you see what's going on with it. A few beekeepers cause problems because they base their consumption on dreams instead of the real world. Anticipate that it should cost more cash than you suspect at the outset and that it will require more investment than assessed to recover your speculation, regardless of representing things to come objectively. For now, expect a few hundred bucks, contingent upon sources and quality, to set up a total hive with two or three honey supers (exceptional hive boxes for gathering honey), with a sovereign and bees included. Currently, it takes a starting capital of something like $300 to start a solitary settlement and buy related hardware. As referenced, having two hives is ideal in the first place, which adds another $200. Bookkeeping One thing numerous beekeepers don't stress toward the start of their profession is monetary record keeping, yet it is fundamental as the venture develops. Current business achievement is predicated on utilizing monetary records and paying proclamations as a long-range system component. Business beekeeping is no exemption. Bumble bees sting, and being stung is a reality of beekeeping life. Stinging is a guarded way of behaving. This can't be said too often, as many people have been molded to consider bumble bees forceful. They are just responsive and protective. Here are a few fundamental focuses on stings. Most frequently, stinging is provoked by conduct the actual bug considers forceful, particularly assuming that the Home seems, by all accounts, to be undermined. Smacking at bees, moving excessively near a home entry, and vibrating the hive or Home are particularly compromising human behavior. On the off chance that a colony has been as of late upset, stinging conduct will doubtlessly increment for a while and may not die down until the following day. Certain individuals are more inclined to be stung than others. This might be because of body science or scent (fragrance, antiperspirant, hair splash) appealing or hostile to bees.

Tell a sidekick if help becomes fundamental. Promptly scratch the sting device out of the skin. Wash the site or potentially apply scouring liquor to lessen the torment. Apply ice to assist with diminishing expansion. Many solutions for stings are casual writing, including tobacco juice, toothpaste, meat tenderizer, and human spit. The reality, in any case, is that the toxin has been infused into the skin, and it is impossible to eliminate it. The main viable medical aid for outrageous foundational responses to bumble bee toxin is infusing adrenaline into the human body. Business "sting packs" containing epinephrine can be found at any pharmacy and normally require a medicine from a doctor.

Every beekeeper should be in the bee yard with such a unit, which frequently incorporates allergy medicine tablets and an adrenaline needle for crises. This goes twofold for anybody hosting human visitors in their backyard. At the point when guests are available, at the very least, foster an arrangement to survey their stung experiences. If side effects of a fundamental response show up, settle casualties with injectable epinephrine found in a sting unit and promptly move them to a trauma center. Bees are enlisted to become protective by a caution pheromone (a substance signal, normally a scent). This important substance has been seriously contemplated and is known as isopentyl acetic acid derivation, which scents like bananas to people. A set-up of different pheromones might be involved too. The justification behind utilizing smoke in settlements is to veil the scent of these substances. Beekeepers become acclimated to stings decently fast and, at times, may not experience a lot of aggravation by any means, contingent upon the site. Stings on the head, face, or fingertips are typically the most ridiculously agonizing. A bee sting, be that as it may, is in every case possibly serious. The seriousness and span of response can differ, starting with one individual, then onto the next and starting with one time, then onto the next.

Moreover, one's response to a bee sting might vary between events. A great many people experience a nearby, non-serious response to bee toxin. Contingent upon the area and several bee stings got, notwithstanding, as well as the consistently present chance of an extremely hypersensitive response to bee toxin, a sting can encourage what is happening. Youngsters frequently experience the ill effects of stings more than grown-ups do. This might be mental or on account of lower body weight. Remember, nonetheless, that youthful people represent things to come in the local beekeeping area. Urge them to look at bumble bees according to the beekeeper's viewpoint — as useful. Bugs, unlike animals, are out trying too hard to find something. Try not to act over the top with a bee sting; put something on the site and give a lot of compassion to the kid. However, at that point, go on as though nothing has occurred. Responses to Stings Two sorts of responses are typically connected with bug stings: neighborhood and foundational.

Sting Reaction Normal Ordinary Response Brightened wheal, focal red spot, confined swelling and torment, dying in minutes to hours. Huge neighborhood responses may likewise happen, dying down in a few days. Interesting Unusual Response Foundational responses now and again show minimal, limited expansion yet can influence organs far off from the sting site. These may bring about hives, inconvenience in breathing due to aviation route expansion, and a drop in pulse. A nearby response is mostly described by torment, enlarging, redness, tingling, and a wheel encompassing the sting site. This is the response of most of the individuals, and this gathering is viewed as a little gamble of death. Numerous in everybody, nonetheless, accept that since they "puff up," they are in danger of losing their life when stung by bees. Unexpectedly, it might truly be told to be the opposite. Those undeniably more in danger might show no impacts from stings by any means until they experience a foundational response, at times called anaphylactic shock. This is most hazardous when the mouth, throat, and aviation routes are impacted, and breath is impeded. Allergists and doctors differ about how to move toward anaphylactic shock (or hypersensitivity). Many favor intense mindfulness and regard most fundamental responses as hazardous. In any case, this predisposition has a little premise, as per Dr. Howard S. Rubenstein, writing in The Lancet: "A considerable lot of the enormous number of individuals who are stung every year by bees experience terrifying fundamental responses," Dr. Rubenstein states, "however by far most of such responses are not hazardous. There is no proof that the few who kick the bucket because of a bee sting come from the pool of the people who once before supported a fundamental response. Going against the norm, no response at all might be a more unfavorable indicator of a deadly result of an ensuing sting." As per Dr. Rubenstein, passing from bee stings comes to fruition through various systems. The most significant is the impact of atherosclerosis (development of stores in the conduits) and unnoticed cardiovascular sickness. Other factors, such as temperature and the sting site, influence mortality.

Gaining Perspective on Bee Stings

Dr. Scott Camazine, writing in the Notice of the Entomological Society of America, says that most have an extraordinary dread toward venomous creatures. In the master plan, he says bug stings are a minor medical condition. Around 40 passings happen yearly due to stinging bugs, most in the request Hymenoptera (insects, bees, and wasps); bumble bees might cause half. Unfavorably susceptible responses to penicillin kill sevenfold the number of individuals, and lightning strikes kill twice as many. Conversely, the country's biggest executioners are cardiovascular infection (100 individuals each hour) and car crashes (one individual like clockwork). Incidentally, Dr. According to Amazing, one is at more the gamble of kicking the bucket in an auto collision en route to the clinic to be treated for a hypersensitive response than biting the dust from the sting that delivered it. Resistance and Treatment of Stings Beekeeper resilience of stings increments with openness.

After several stings, many acknowledge they are no huge thing and overlook them when they happen. One of the creators once saw the grayness of one encountered beekeeper's hand at a bustling apiary. On nearer assessment, it became clear that his hand was, in a real sense, concealed with dried stingers that he had not tried to eliminate. Over the long run, the human body becomes acquainted with the toxin. The consequences of stings as far as torment and enlarging are, in this manner, frequently diminished. The response to stings might move over the beekeeping season, being more limited toward the start when beekeepers initially start to visit states. For some's purposes, in any case, resistance to stings can take an alternate course. These people respond more with each progressive sting and can foster a genuine sensitivity needing clinical help. This can happen to beekeepers as well as non-beekeepers. It has been noticed most intently in beekeepers' families, particularly those not dynamic in the specialty. The rehashed breathing of particles of dried toxin or other bee-related materials that stick to beekeepers' clothing can sharpen other relatives. It is, in this manner, suggested that beekeeping clothing be washed as often as possible and that it and other beekeeping gear be set where others are not presented to it.

Stings by Africanized Bees

The African bumble bee was brought to Brazil during the 1950s, and its intersection with the generally settled European bee brought about a crossbreed known as the Africanized bumble bee. This bug slowly penetrated the western US and is also present in Florida. Even though sickness is lenient and a decent brood-maker, the Africanized bumble bee has gained notoriety for unreasonable amassing and protectiveness. Since the 1970s, sensationalized reports of its stinging behavior have besieged the population. Dr. Mark L. Winston, in his book Africanized Bees: The Africanized Bumble Bee in the Americas (1992), considered it the "pop bug" of the 20th hundred years. Different stings might bring about uncommon clinical issues like encephalitis, polyneuritis (degenerative or provocative injuries of a few nerves simultaneously), and renal disappointment. The last issue has happened because of mass bee assaults in Latin America by Africanized bumble bees, something else entirely from episodes including only one or a couple of bees. Any individual paying little mind to aversion to bee toxin, getting a gigantic number of stings (mass envenomation), may be vulnerable to renal disappointment or other extreme issues because an incredible amount of poison overpowered their body.

Any individual paying little mind to aversion to bee toxin, getting a tremendous number of stings (mass envenomation), may be helpless to renal disappointment or other serious side effects essentially because an incredible amount of poison tested their body. Where Africanized bumble bees are laid out, doctors ought to retrain themselves for the last chance, where the medical aid method is renal dialysis, not injectable epinephrine. Where Africanized bumble bees are laid out, doctors ought to set themselves up to treat with renal dialysis instead of injectable epinephrine. The deadly portion of bumble bee toxin in people is around 19 stings for every 2.2 pounds (1 kg) of body weight or around 1,300 stings for a 150-pound (68-kg) individual. If you have been mulling over everything, do think about attempting it. The primary year is a major expectation to learn and adapt. However, it is worth the effort presumably happens in under 1% of the populace and just a little level of those with a sensitivity foster extreme responses. Indeed, even with the appearance of the Africanized bumble bee and related stinging episodes, Dr. Camazine finishes up, there is no obvious explanation to think that bee stings will become a huge well-being peril. There is more motivation to be worried in regions where Africanized bumble bees are laid out, yet the chances of being gone after are tiny. The Africanized bumble bee keeps on turning out to be more dug in North America, notwithstanding, so it pays to know where populaces are laid out. In any case, less than 30 passings from stinging occurrences have been kept in the US, where this bee has been laid out since its presentation in 1990. Impact of the Sting on the Bee Assuming there is any reassurance, when the bumble bee stings, she surrenders her life. The bee doesn't bite the dust right away: she can and frequently keeps irritating an apparent danger and select others for a similar movement. The mechanical sting assembly stays in the person in question, notwithstanding, torn from the bee's body, and the bug is harmed hopelessly and kicks the bucket. Scientists have seriously researched the wellspring of this benevolent conduct over numerous years. It entrances many individuals and is one reason they become inspired by friendly bugs, maybe eventually taking up beekeeping.

CHAPTER 2

THE ORIGINS AND EVOLUTION OF BEEKEEPING

Keeping Bumblebees with a profound knowledge of the bug's science or background IS Conceivable. Even so, you will be a substantially more compelling director if you figure out the elements of the bug and its relationship to the colony, different creatures, and the climate. Beekeepers keep a single type of bug, the Western bumble bee known as Apis mellifera, signifying "honeycarrier" or "honeybearer." There are many types of bees, in any case, and a lesser but no less significant number of related wasps. The population will generally toss all stinging bugs into a typical pot called "bees." Even though bees and wasps might seem comparative, they are not. Bees are vegans, polishing off sweet plant juice called nectar (starch) and dust (protein). Wasps likewise guzzle nectar, yet they take their protein from different creatures; they are predatory and frequently go after different bugs that may be human or bee bothers. The two gatherings are useful bugs and should be secured, albeit numerous people need help understanding this.

APIS (Bumble Bee) Genealogical Record

Bees and wasps share the accompanying order:

Realm. Creature Phylum. Arthropoda: bugs, insects, shellfish Class. Hexapoda (six	or Insect Request. Hymenoptera (film winged): bees, wasps, subterranean insects, sawflies, horntails Suborder. Apocrita (alluding to restricting or choking of the mid	bees, wasps, and insects. Bees veer from wasps at the superfamily level: Superfamily. Apo idea: bees. Wasps and insects are in the Vespoidea superfamily Sort. Apis: bumble bees Species. Mellifera: Western bumble bee

Wasps and bumble bees are firmly related: as a matter of fact, bumble bees are believed to be the relatives of tissue-eating wasps. The accompanying advances were significant in developing the present-day bumble bee society, which keeps showing these ways of behaving: Individual taking care of. A solitary, crude mass taking care of the youthful, as polished by numerous bugs, gave way to moderate taking care of it, furnishing the youthful with predigested food when required. Shared society. Continuously the sovereign or conceptive individual and her little girls started to live respectively, and the colony became perpetual with an obvious position system by which every individual played a specific part in the Settlement. Complex correspondence. Correspondence among people developed into a perplexing system, including food sharing (trophallaxis), compound exchange through pheromones, and elaborates moves. The Western bumble bee, Apis mellifera, is a solitary animal variety that has moved and been shipped by people all over the planet. When generally remembered to have begun in Asia, ongoing DNA examination proposes that the species emerged in Africa. It is remembered to have moved from that point by two courses: toward the west through North Africa and the Iberian Promontory into Europe and toward the east through the Center East to Asia and Europe. Bumble bee Ecotypes Throughout the long term, bumble bees have arranged themselves into various subspecies or races of discrete populaces in light of neighborhood climatic circumstances and nearby plants they can use as nectar and dust sources. These subspecies are called ecotypes. Thirteen exist in Africa, six in the Center East and Asia, and around nine in Europe. Those brought to the New World have generally been European, including the German or dark bumble bee. Apis mellifera, the Italian bumble bee; A. mellifera ligustica; the Carniolan bumble bee; A. mellifera carnica, the Caucasian bumble bee; and A. mellifera caucasica. Non-European acquaintances with the Americas include the Egyptian bumble bee, A. mellifera lamarckii, and the Syrian bumble bee. A. mellifera syriaca, the African bumble bee, and A. mellifera scutellata. These ecotypes can interbreed, implying that half and halves can likewise be found.

Bumble Bee

Wasp

 Many distributed data portray contrasts among the bumble bee races currently viewed worldwide. Italian bees (Apis mellifera ligustica) produce a lot of brood all year, are extremely peaceful on the brushes, and are somewhat more impervious to American foulbrood than others. Carniolans (A. mellifera carnica) quickly change their brood raising to the season and have medium-length tongues. The short-tongued German or dark bee (A. mellifera) is usually cautious and powerless to illness, while the Caucasian (A. mellifera caucasica), with the longest tongue, is inclined to foulbrood contamination and gathers significantly more propolis than different races. Throughout beekeeping in the US, the four significant European races of presented bees referenced above have, as has the human populace, lost their singular character and vanished into an extraordinary blend. A reasonable guidance plan has been expounded on boosting honey creation, controlling amassing, and other administration rehearses throughout the long term. In the whirlwind, in any case, the dominating race of bees being kept is frequently overlooked. That is reasonable because it is hard to coax out unambiguous gatherings that occupy a specific settlement. Giving guidance or going with the board choices that rebate the way that all bees are different can be counterproductive. It isn't the capacity to respond to the situation with individual hives yet rather dealing with the fluctuation among settlements that better characterizes the genuine bee ace. Bumble bees are social bugs. This intends that even though people are the reason for any colony, they can only exist in detachment with assistance from different individuals. The colony's productivity makes this bug so fruitful in any place it is found, and this is the reason for its administration by people. Beekeepers like this don't oversee single bees.

A Thorough Historical Overview

Most beekeeping innovations were created during the 1800s; instrumental insemination is the main procedure considered an offspring of the 20th century. The historical backdrop of apiculture here and there matches that of science itself. From ancient times through the Dull and Medieval times, little was some significant awareness of bumble bee science. A stone composition dating from around 6,000 BC in Cueva de las Arañas, Spain, is replicated in numerous beekeeping books to address ordinary beekeeping movements before current times. It shows stick figures scaling stepping stools and eliminating wild homes. Honey hunting or home evacuation goes on in Asia with the monster bumble bee, A. dorsata, which has never been brought under beekeeper the executives. Throughout the long term, creating information on bee science impacted the specialty in numerous ways. Dynamic administration of the bumble bee colony, in any case, anticipated key snippets of data that arose in the eighteenth and nineteenth hundreds years. In later times, developments have expanded as headways in transportation, correspondence, and different advancements have become possibly the most important factor.

At last, the new natural objectives of a worldwide local area and economy have impacted beekeepers in manners not longed for by pioneers. A few periods are depicted in this volume describing the improvement of beekeeping: Ancient times to 1500; 1500 to 1850, 1850 to 1984; and 1984 to right now. Ancient times to 1500: Home Burglarizing and HoneyChasing After hundreds of years, bees were kept in empty logs or straw containers. Before 1500, the level of the Renaissance, beekeeping comprised minimal more than ransacking honey from laid out homes. The Philistines fiddled with beekeeping, as did the antiquated Sumerians, Arabians, Greeks, Romans, and Egyptians. Beekeeping methods included hiving the bees and afterward gathering the honey through the Home obliteration. Bees were housed in various homes, from empty tree trunks (gums) to earthen stoneware to woven straw containers (skeps). The bugs were urged to repeat by amassing because this was the best way to populate new homes given by the beekeeper. The bumble bee isn't local to the New World, so Native Americans have no set of experiences with this bug. The Mayan and other New World developments kept certain tropical or stingless bees, notwithstanding some methods like those utilized by European beekeepers.

Bee lining Method

Beelining is a technique for finding wild homes that exploit the bumble bee's normal behavior. Initially, a few trap stations are set up to draw individual search bees. After taking care of, these bugs fly in an orderly fashion (a beeline) back to their Home, enlisting home mates to rummage at a similar spot. In the drawing underneath, we can undoubtedly see the "secret hive" through triangulation by defining crossing boundaries from the lure stations. Beelining is as yet utilized in certain areas of the planet. Most outstanding is the ongoing exertion in Australia to track down homes of wild Africanized (Apis mellifera scutellata) and Asian (A. cerana) bees that have escaped at ports. Contemporary observation officials Down Under have removed a page from the beelined's manual to guarantee that these exotics don't become laid out in one of the world's head honey-creating countries.

The Globetrotting Bumble bee

As an acknowledgment of apiculture as a real work proceeded, the bumble bee spread overall with beekeepers' assistance. This goes on even today. Huge occasions and dates include the 1860s: Italian bumble bees were previously brought into the US. The 1870s: Straight to the point, Benton brings Cyprian and Tunisian stock into the US. 1957: African bumble bees are brought to Brazil. The resultant wild bees (before long named Africanized bumble bees) soak northern South America, and later Focal America lastly arrived in Texas in 1990. The 1970s and 1980s: Honey turns into a world product. The 1990s to the current day: Business fertilization turns into an undertaking that permits business beekeepers to settle their pay, and most, as of late, have caused a blast in almond creation in California. Annihilation drives for both these bugs would be ineffectual. It is no misrepresentation to say that a condition of close to overreact impacted bumble bee specialists and beekeepers as these irritations spread to almost every bee yard in the landmass. Beekeepers, with an enemy of pesticide inclination, embraced when crop ranchers' utilization of these materials had killed many of their states and immediately understood that they should now utilize them to get by. This was particularly the situation with the Varroa vermin invasion; on the off chance that not treated, it would rapidly kill states because the bugs had no natural obstruction or resilience.

The perception hive is a debut research and instructive device for beekeepers. It can likewise be utilized as an assistant to various advertising and selling programs. Even though its charm is widespread, the perception of a beehive may differ from the ideal display decision. A significant investment is expected to create a hive and make a big difference. Most people have a few issues introducing a perception hive. The following cerebral pain is keeping up with the unit. This is particularly obvious if the hive is utilized as an extremely durable showcase for the overall population. Sadly, almost no that is long-lasting about a perception beehive without extensive work by the beekeeper. Indeed, even the biggest units of the four approaches still address a piece of a standard state. Since they are so few, perception hives don't generally endure significant variances in either populace size or food accessibility. Any individual endeavoring to save one of these negligible states for any period can draw up a long clothing rundown of likely issues. These can include amassing, queen-lessness, starvation, and attack by infections, nuisances, and parasites. It is not necessarily the case that the perception hive doesn't have a spot, just that a guarantee to oversee it should be made for a while. Nothing is more regrettable for a public showcase than a disregarded perception beehive.

CHAPTER 3

Acknowledging a Bee's Life

HONEYBEES are wild Animals. They have never been trained. They have been kept, examined, explored, and reproduced for a long time, and, as it were, the species has gotten to the next level. Be that as it may, they have not been restrained. Just let alone, they live precisely as they have lived for millennia. Our valid, long-haul accomplishment as beekeepers comes solely after we come to grasp their private lives, conduct, and inspirations.

What Is a Honeybee?

The honeybee has been known as a "flying Swiss Armed force blade" in light of its mind-boggling structure. It is a genuine bug with six legs, three body parts (head, chest, mid-region), a hard external skeleton (exoskeleton), two sets of wings, an open circulatory system, a tracheal breathing system, and a ventral sensory system. The honeybee's head contains two receiving wires, two compounds, and three basic eyes (ocelli). Compound eyes identify captivated daylight, example, and variety, while straightforward eyes uncover enlightenment force (splendor). Mouthparts incorporate matched jaws (mandibles and maxillae) and a solitary tongue (labium) that make up the sucking contraption of the honeybee. The chest (the center part of the body) oversees honeybee development. It holds the enormous aberrant flight muscles to which the wings are joined. Six legs (three sets) likewise interface with the chest. The mid-region contains various organs and systems.

Stomach related System

The stomach-related System comprises the mouth, throat, honey stomach (the region where nectar is put away and shipped), proventricular valve and proventriculus, front digestive tract (ventriculus), and rectum. The throat leads from the mouth to the ventriculus and discharges into an inflatable design called the honey stomach or the "crop." This is a field specialist honeybee's freight hold and can depend on 85% of her weight when full. This nectar never goes into the remainder of the gastrointestinal system, yet it comes by the ventricular valve. When a forager returns to the colony, the items in the honey stomach are spewed through the mouth and given to the house bees for additional handling. This implies that nectar appears extremely near its regular state as it was gathered in the field and thus contains no conceivable tainting from the remainder of the gastrointestinal system.

Circulatory System

A honeybee has a dorsal heart yet not many veins. This is known as an "open" system because the blood sloshes around inside the bug's fingernail skin or skin. The blood, called hemolymph, is different from hemoglobin (a protein in human blood) since it doesn't convey oxygen, which should be obtained straightforwardly from the air. (See conversation of respiratory System beneath.) Respiratory System Cylinders or tracheae directly trade oxygen between the body tissues and the external air through outside openings (ostia). The tracheal System may be most popular among beekeepers since it can uphold a populace of vermin, which can be unsafe in high focuses.

Excretory System

This comprises little cylinders, called Malpighian tubules, that float in the honeybee's hemolymph, gathering pollutants kept in the rectum.

Sensory system Various ganglia (structures containing nerve cells) make up a focal sensory system that starts in the head and then courses along the lower body. This ventral System is portrayed by matched nerve strings significant in directing muscles and organs. A Broad Organ Complex incorporates hypopharyngeal (brood food), mandibular, wax, and fragrance organs. Different organs are tracked down in many bodily puts, including the feet, the mechanical string assembly, and the skin (fingernail skin). Inside the Settlement, As noted somewhere else in this volume, the honeybee colony is appropriately called a superorganism. It is eusocial — a gathering of creatures acting together, portrayed by helpful consideration of the youthful, covering ages, and dividing work through a standing system. Most eusocial life forms are bugs like subterranean insects, bees, wasps, and termites, yet a few are found in different gatherings, for example, the bare mole rodent, a well-evolved creature. The Home In the wild or the hive, the honeybee home comprises equal brushes loaded with hexagonally molded cells. The hexagon is the primary structure that gives the most strength utilizing a minimal measure of material; a demonstrated numerical maxim. The brush is developed exclusively of beeswax, created by the honeybee's body, and fills a double need as both the storage room for the colony's food and the support for the youthful bees (brood). The focal comb(s) typically have a particular example, with the youthful bees or potentially brood encompassed by a band of dust and, afterward, honey. As the brushes are worked out from the Home's focal point, not so much brood but rather more dust and honey are often viewed as on each. The side brushes, by and large, are brimming with honey and are brilliant separators, keeping up with the intensity important to keep the brood creating.

Particularly critical for the colony is how the honeybee transforms completely. Every station goes through a formative cycle where the body fundamentally changes structure and capability. Starting with an egg, which hatches into a little worm (hatchling), the singular moves into a resting stage called the pupa. During the pupal cycle, tissues are adjusted into the last stage, the grown-up. Accordingly, the singular honeybee has explicit ways of life: eating and creating (the egg, hatchling, and pupa, altogether called brood); and adding to the state's support and development (the grown-up). Advancement happens in the phones of the brush. When an egg comes forth, the workers initially give it a nutritious jam (imperial jam) in a cycle called mass taking care of. As the hatchling creates, the grown-ups give food depending on the situation, a cycle called moderate taking care of. Toward the finish of the larval period, the bees cap the phone with wax, and the hatchling then, at that point, rearranges its tissues through a cycle called pupation. When this change or transformation happens, the full-fledged grown-up bites away the cap of its cell and joins different grown-ups in the colony. The Queen, The queen is the conceptive focal point of the state. She is the main female equipped for laying prolific eggs, which the state needed to fill in the populace. She is likewise the longest-lived individual, once in a while enduring quite a long while, and the biggest, with a long, tightening body. Since her responsibility is to lay fruitful eggs, the queen has around 160 ovarioles, where the eggs structure (workers have four and no more), and a solitary spermatheca, where sperm is put away. Albeit the queen comes up short on qualities of the workers (sterile females) in the state, like wax organs, dust bushels, and a fragrance organ, she is as yet ready to sting. This helps her control equals. However, it must be utilized in everyday protection. People are seldom stung by queens; however, assuming they are, the stinging device doesn't stay in the person in question (not at all like with workers) because its points are tiny. Queens likewise have scored jaws or mandibles, which separates them from workers' smooth ones. There gives off an impression of being a formative continuum among "queendom" and "workerdom." Queens who look more like workers are called intercaste. These are interesting, and it has yet to be known when or how a state manages them. The queen's turn of events or transformation is short when contrasted with that of others in the colony.

The egg hatches in 3 days, the hatchling pupates in a little more than 5, and a grown-up arises seven days after the fact. The complete improvement time is 15.5 days. After a couple of long periods of development, the queen is prepared to fly and start the method involved with mating, securing the sperm essential to the colony's future. In the week or so after first taking wing, she should situate herself to the colony and afterward make a few trips during which she is sought after and mated with by upwards of 17 to 20 drones. When she starts to lay rich eggs, she will leave the colony again with a multitude when the bees move to lay out another home.

Queen Impression

Most researchers don't think the queen "runs the show" a settlement, yet she positively applies control. The main errand is to utilize synthetic compounds, or pheromones called queen substances, to keep workers' ovaries from creating where they produce eggs. Workers will lay eggs every so often —for example, when the queen kicks the bucket out of nowhere without a substitution — however laborer delivered eggs can't be prepared, resulting only in drones that can't support a settlement. Since no workers create to supplant workers who kick the bucket, the colony quickly lost the populace. Another way the queen applies control is by deciding the sex of her posterity, creating either drone from unfertilized eggs or workers from prepared eggs, so she is answerable for the possible populace cosmetics of the Settlement. The component is surely known — she either does or doesn't deliver sperm as an egg passes down her regenerative parcel — yet the way things are achieved and under what conditions isn't known. Since she has mated with a huge number, the queen's posterity comprises various subfamilies of workers, some more connected (super sisters with a similar dad) than others (stepsisters with various dads). Egg creation by the queen's rhythmic movements with state improvement depends on many elements, including winning weather patterns, wholesome assets, and the cosmetics of the populace inside the hive. In the dynamic season, a few queens may rest up to 2,000 eggs daily if a huge populace is required; however, when the circumstances change, egg-laying does too. There is normally just a single queen in a colony, yet under certain conditions, at least two may be found. The standard clarification is that a mother and a little girl — the girl who will supplant the mother — are available briefly on a similar brush. Various virgin queens might sometimes appear in swarms, particularly among Africanized honeybees.

The Worker

The specialist honeybee is a full-grown female, yet the impact of the queen's pheromones (queen substance) keeps her ovaries little and lacking. The specialist populace is answerable regarding the very aware honeybee exercises, including honey creation, dust assortment, temperature guidelines, brood raising, wax and regal jam creation, and safeguard. There are more workers in a settlement than in some other position, numbering several thousand. Specialist honeybee Workers are the littlest people in a settlement and typically have the briefest life expectancy, maybe enduring half a month in the dynamic season. The specialist might endure a few months in the idle season, particularly in calm locales. They are additionally the most excessive, wearing themselves out, keeping up with the Settlement, focusing on the brood and rummaging for food, or surrendering their lives with all due respect. The specialist starts life as a prepared egg laid by the queen and brings forth into a hatchling in 3 days. The hatchling is taken care of for about seven days and afterward pupates in 12 additional days, arising as a grown-up following a sum of 21 days. The 3-week improvement pattern of the specialist is essential to the beekeeper endeavoring to screen a colony's populace, either for honey creation or fertilization.

By and large, most house bees go through several errands. Probably the most huge are recorded here and show up in the rough request of when they can be achieved formatively: Cleaning brood cells Going to queens Taking care of brood Covering cells Pressing and handling dust Discharging wax Directing temperature Getting and handling nectar into honey There are both colony and hereditary variables engaged with figuring out what workers do as they age. Tending the Nursery A few specialist errands require the utilization of explicit organs. The house bees feed the hatchlings a nutritious jam called brood food, which is delivered in the hypopharyngeal organ in the honeybee's head and, in analysis, looks like a lot of grapes. The organ is useful for a brief time frame during the house honeybee's initial improvement when it is generally required. Be that as it may, if the Settlement decides this brood food is important and more established workers are accessible, they can recover the utilization of the design. Brush Building Late in the existence of a house honeybee, the wax organs become noticeable, arriving at a top being developed at around 18 days. They show up as four discrete sets of wax mirrors on the underside of the honeybee's gut. The organs discharge a fluid that then, at that point, cement on the mirrors. The honeybee then eliminates these strong wax drops and forms them into the brush. Once more, albeit more established, working drones don't have totally useful wax organs; they can reactivate them amid settlement needs. Honey Processors House bees are the superb honey processors in a settlement. They get crude nectar from rummaging bees and spot it in the brush where vanishing happens. This is done by dividing the sweet between themselves by trading drops of nectar and storing them in different cells around the Home. Nectar at this stage is extremely similar to water apparently and consistent. The workers then fan their wings to carry dry air into the colony and exhaust it, brimming with dampness, to the outside. During this period, the bees add compounds that artificially change the nectar. These compound changes, combined with a decrease of dampness in the nectar, bring about the item known as honey.

The Drone

A little while after the rise, the house honeybee starts to fly from the colony, planning to turn into a forager. This direction or "play" flights are superb to see. The youthful bees frequently arise in a cloud on a warm day and start to fly volatile before the hive, gradually voyaging increasingly far from the Settlement until they realize the encompassing scene adequately to return. Even with this direction period, be that as it may, numerous bees effectively get lost, particularly assuming various hives nearby. This "floating" conduct isn't an issue in the wild, where states are generally far separated; however, in an apiary with various hives, working drones frequently become confounded and may go into different settlements. This produces hives of inconsistent size and can likewise spread illnesses and vermin throughout the apiary. Scrounging isn't arbitrary for field bees: They are generally directed by supposed scouts to the best nearby nectar sources. Field bees are often devoted to this plant, visiting it solely when they start scavenging on a specific yield. This makes them effective and positive pollinators. Scavenging is risky. Numerous animals, from ruthless wasps to bugs, successive plants in sprout. Openness to climate components and poisonous pesticides is dependably a chance. The general view is that searching bees essentially wear themselves out after some time. This is much of the time observable in more seasoned foragers. In the long run, their wings become worn out and less useful.

The Winter Honeybee

The existence of the working drone relies upon climatic circumstances. In calm districts with unmistakable winters, there are two sorts of workers: "summer" bees and "winter" bees. Structures called fat bodies in winter bees empower them to get by during chilly climates when settlements produce nothing, brood or food. A state's creation of winter bees is critical, and expanding this populace becomes a significant administration methodology in northern regions. Fat bodies are absent in summer bees. Researchers need to comprehend how the proportion between summer bees and more winter-adjusted not entirely set in stone in the subtropics and other less calm locales, yet it is no question a significant result of the different subspecies or ecotypes that have been created in unambiguous regions. The Drone

The drone is the male honeybee, a surprising life form by his own doing. He, most importantly, is haploid, the consequence of parthenogenesis or "virgin birth." That implies he has a portion of the number of chromosomes as the female. He is the perfect hereditary representation of his mom (and, in this manner, has no Y chromosomes) and has no dad. His regenerative System is intricate and comprises matched testicles, vas deferens, original vesicles, and mucous organs, as well as a solitary reformatory bulb and related structures. The drone can be viewed as what might be compared to a flying sperm. Most researchers and beekeepers accept his main responsibility is to mate with a queen, furnishing her with many duplicates of himself in the deal. It is conceivable, in any case, that drones carry out different roles in the hive, including temperature guidelines. The most common way of limiting or dispensing with the drone populace could have unseen side effects that beekeepers or scientists do not perceive. The drone kicks the bucket in the mating system: the breeze pressure created during his endeavor to catch and mount the queen makes him detonate with a discernible pop as he discharges inside her. He then, at that point, tumbles off the queen, typically leaving a part of his phallus inside her. Research on drones uncovers that they could be more consistent and scattered across the scene. Explicit assembly regions contain assortments of guys from all through the area who structure more modest units (called comets) and seek after individual queens. Unlike workers, drones are acknowledged in all states at practically any time. The drone has the longest formative time of any station. Like the queen and specialist eggs, the drone egg hatches in 3 days. The larval stage is 6.5 days, followed by 14.5 long stretches of pupation, adding up to 24 days. This more drawn-out improvement time makes drone brood more appealing to the Varroa parasite; one compelling biomechanical control involves catching bugs in drone brood and eliminating it — and the vermin alongside it — from the colony. At most, a couple of thousand drones live in a state (which can number 60-100,000 bees) during the dynamic season.

Researchers can now prepare honeybees to distinguish one-of-a-kind substance particles. A wonderful 95 percent precision rate, contrasted and 71 percent in bomb-sniffing canines, has opened up the possibility of utilizing bees to recognize Ad libbed Unstable Gadgets (IEDs), self-destruction planes, organic weapons, and ultimately illnesses and diseases in individuals. Most refreshing is the genuine chance of utilizing honeybees to identify the numerous plastic hidden explosives that populate the earth and cause a lot of human misery. The bees' preparation utilizes old-style Pavlovian conduct molding (zeroing in on their proboscis expansion reaction), and progressively more remarkable and affordable procedures distinguish and examine the outcomes. At least one gadget interprets honeybee conduct into an electronic sign that can be distinguished on a screen. Then again, honeybees can learn. Different analyses have demonstrated this, many utilizing a similar standard of social molding (learning new conduct because of explicit experience, positive or negative) that Russian researcher Ivan Pavlov utilized with canines. In honeybees, Pavlovian social molding is effortlessly seen in their proboscis extension response (PER), and they have been prepared to perceive a wide range of substances, from medications to explosives.

Honeybee Feelings Whether honeybees experience misery isn't known, even though there is an Old World custom of "telling the bees" when their beekeeper has passed on. Furthermore, even though it is impossible to determine what bees themselves feel, that's what some trust; as canines, bees can identify apprehension in a beekeeper. It is many times said that honeybees can perceive their beekeeper. Design acknowledgment has been identified in honeybees; it is conceivable that they become sensitive to a particular human face. It is improbable that singular bees get to know any individual because the typical beekeeper doesn't visit a particular hive frequently enough. A singular specialist honeybee might live simply for 4 to 5 weeks in the dynamic season, so there isn't much chance for openness to a solitary person. More probable, the bees center around a human beekeeper's behavior. Another chance is that science is involved. Many individuals draw in light of a legitimate concern for bees since they regularly wear fragrances, scented antiperspirants, or styling. A few creatures, like ponies and canines, are known to evoke cautious conduct in honeybees, definitely more than different organic entities. It is thought this is the consequence of specific scents, ways of behaving, or a mix of the two improvements. Examples of Conduct Honeybees show numerous personal conduct standards that beekeepers should consider while overseeing settlements. A model is the state's response to smoke, a significant beekeeping device. When a state is effectively smoked, bees first create some distance from the source and afterward engorge themselves with honey from the brush, making them less inclined to sting. Other designed ways of behaving incorporate temperature guidelines (thermoregulation); amassing, which is important for the regenerative cycle; and fleeing, or relocation, in hotter environments. These straightforwardly influence the prosperity of a colony and its efficiency. The best beekeeper perceives these examples, grasps their importance, and stands prepared to respond if it would be a good idea for it to become fundamental.

Temperature Guidelines

Unlike well-evolved creatures, individual bugs are heartless and can't control their internal heat level. In this manner, the thermoregulation of social bug colony is an extraordinary development in the method for surviving with the accompanying benefits: A honeybee settlement can be dynamic all year, in any event, when temperatures are outrageous. In mild regions, honeybees can deliver more brood before the season than bees with yearly life cycles (honeybees). The honeybee could see both intensities and cool its Home, keeping the brood at any point home temperature at a predictable 95°F (35°C). Bunching for Warmth During chilly climates, individual laborer honeybees shiver or vibrate their flight muscles to make heat. They do this while crouched together in a ball or "group." The more seasoned bees line, the beyond the bunch to protect it from heat misfortune. More youthful bees are inactively warmed inside the mass of bees. A precarious temperature slope frequently shapes between the bunch's middle and the outside, and the wad of bees extends and contracts as the bees draw farther separated or nearer together because of the encompassing air temperature. The bunch of bees structures when the temperature outside decreases to around 57°F (14°C) and disassociates at higher temperatures when heat creation is excessive for the hive.

Cooling Honeybees cool the Settlement by gathering water and fanning air flows through the state. During a sweltering climate, an enormous number of bees may likewise balance outside the hive around evening time, appearing to dribble off the base load up. This mass of workers is known as facial hair growth, and the way of behaving is called bearding. Even though fledglings frequently dread this, it indicates that amassing is unavoidable; that isn't true; it is an ordinary reaction to high temperatures

Chapter 4

Selecting a Location for Your Beehive

FOR BEEKEEPERS, Concerning Land Specialists, a large part of the time achievement reduces to "area, area, area." It is likewise well to remember one more beekeeper's proverb, summarized from the political domain: "All beekeeping is neighborhood." Figuring out where to find a hive is frequently clear in light of limits; however, numerous choices might be conceivable in different circumstances. No matter what, it is ideal to see expected destinations according to the bees' perspective. Concentrates on finishing in mild areas, where homes are typically tracked down in empty trees, uncover specific normal inclinations. The bugs pick a raised area, around 10 feet (3.0 m) off the ground, with a southerly openness and little sun or shade. In a few tropical circumstances, be that as it may, trees are not accessible, and bees might need to make do with sub-par conditions. Besides the bees' inclinations, the beekeeper should consider openness and the possible misgivings of neighbors, meanwhile applying a proportion of sound judgment. Four Hints on Hive Area Out of the picture and therefore irrelevant. The nearer your hives are to your home, the better. Amateurs need states close by so they can visit them frequently. Relaxed visits are significant regardless of whether you open the hives. You can learn a lot just by noticing the entry. Try not to welcome defacing. Disconnected beehives can draw into consideration of hoodlums, transients, exhausted children, and cheats. States have been taken completely, pelted with rocks, spilled, designated with gunfire, and thrown into waterways. Keep them where you can see them effectively; however, cover your hives behind the growth and additionally find them behind walls and structures.

Roof Hives

One of the most incredible areas for honeybees is a rooftop. Numerous apiaries flourish with city housetops from Paris to Chicago. The advantages of a housetop area include how your hives stay away from obstruction by people and different creatures. What's more, they are relatively easy to screen. Metropolitan bees scrounge in parks, gardens, road trees, and overhang blossom boxes: any nectar source inside around 1 mile (1.6 km) of the state is possibly accessible. City bees now and again can be more useful than country bees. The difficulties of keeping bees on the rooftop incorporate the trouble of lifting honey-filled supers manually since you can't utilize trucks or other huge mechanical gadgets. Furthermore, a rooftop can be boiling in summer or potentially very cool in winter when you'd prefer not to be out on the rooftop yourself! Luckily, bees are great at keeping up with their temperature. Be certain the covers are solidly held and set up by cinderblocks or different loads in the event of high wind. Give a space to work. It takes space to control settlements, so don't find them excessively near each other. Continuously guarantee adequate room between settlements to be worked without upsetting those nearby. Remain level. States should be on level ground. Honeys weighty, and moving it is, in many cases, no tomfoolery; the beekeeper ought to utilize helps like wheeled carts, hand trucks on slanted planes, and mechanical lifts to ship filled supers.

By laying out only one beehive, you present many bugs in your local area. This will undoubtedly have a natural effect, regardless of whether unpretentious or harmless. It means quite a bit to contemplate the impact on your area's different occupants, humans and creatures. Consider the Neighbors Metropolitan beekeeping brings into center various difficulties. Neighbors are a consistent worry for beekeepers. The fledgling should try not to turn into the casualty of the people who consider any bug a disturbance, particularly one that stings. This goes in spades for anybody keeping bees where Africanized honeybees have become laid out. Next are a few recommended strategies to avoid potential issues brought about by beekeeping action.

Bees in provincial regions deliver less than those in metropolitan and rural regions. During the 70s and 80s, I kept exact records and arrived at the midpoint of 30 pounds of honey (13.6 kg) per country settlement, including all that had bees. I kept one hive in my folks' yard in a rural lodging advancement, and it was an uncommon year that it didn't create 100 pounds (45 kg) of excess honey. I am curious to know how much nectar creation has moved as far as plant species utilized, yet where the honeys delivered is obviously in the areas where individuals reside. Charge Starrett, Xenia, Ohio. Look into nearby guidelines managing keeping bees or any creatures in your area or region. On the off chance that there are grievous limitations, consider reviewing a particular law for your circumstance. Numerous neighborhood and express beekeepers' affiliations will help in such a manner, and a few decent model statutes are likewise accessible. Spot states from part lines and involved structures. If close to structures, find hives from doorways and lines of people strolling through. Hive passageways ought to confront away from intensely voyaged regions.

Supplying Water

Numerous regions where bees are found might encounter dry times over the year. When irregular rivulets quit streaming and treeing leaves indicate dampness stress, bees become more recognizable to the overall population. This can amount to calls about honeybees gathering water from spilling spigots, water basins, pet dishes, and pools. The beekeeper should give a water source to bees on the off chance that there is any probability the bugs will scavenge in neighboring metropolitan regions during droughts. Counteraction is the main solution for this issue. Try not to allow the bees to become prepared for a watering place like a pool. When a water scavenging design has been set, successfully transforming it isn't easy. Finding bees close to open water is the ideal way to give a consistent inventory. It is additionally vital to ensure that any potential water supply isn't tainted. If no source is found nearby, giving water in the apiary is conceivable; however, it frequently requires a reasonable setup of arranging and thought. Fifty-five-gallon barrels or different compartments can be loaded with water and layered on top of wood floats to keep the bees from suffocating. An issue with this sort of gadget is possible stagnation. Presumably, the best gadget streams water down a wooden board or gradually dribble onto a porous material, keeping the surface moist. Erect a 6-foot (1.8 m) blockade between the hive and the parcel line.

Use anything bees won't go through, like thick bushes and fencing. There are likely issues whenever bees fly near the ground and across a neighbor's property line. Give a watering source. Place a tub of water in the apiary if a characteristic water source isn't found close by, particularly when pools are nearby. Add wood floats to keep the bees from suffocating. Change the water occasionally to keep it away from stagnation and mosquito reproduction. Honeybees frequently drink from chlorinated pools, which won't hurt them yet can upset people nearby. Honeybees have been prepared to utilize metropolitan watering sources by adding rejuvenating balm or light scents. Limit looting by other honeybees. Looting bees are generally very protective and will be bound to sting bystanders. Control the hives just during nectar streams, if conceivable. Keep extra honey or sugar water from being uncovered, or looting will probably start. (See box inverse.) Use entrance minimizers to limit the probability of more grounded states burglarizing more vulnerable ones.

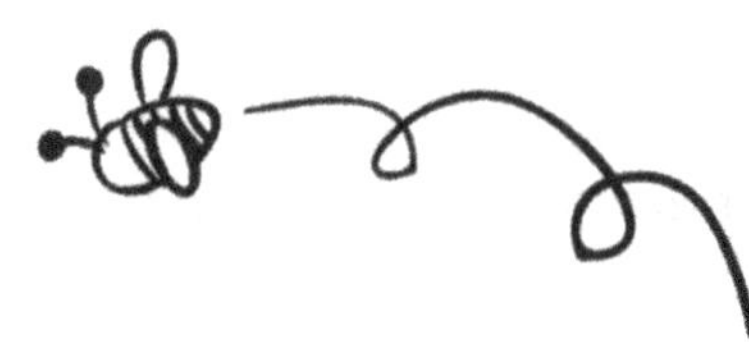

Burglarizing

The beekeeper should consistently prepare for looting for various reasons. Most significant is that this is a significant way bacterial illnesses are spread. Looting can likewise prompt serious stinging episodes. The fervor level shown by honeybees that took part in an enormous burglarizing episode should be addressed. Entire apiaries have been supposedly obliterated on certain occasions. Some burglarizing presumably continues constantly among colonys. This endemic looting level turns out to be just pestilence when conditions are correct. Novices frequently accidentally permit looting to develop when they keep states open for significant stretches while making examinations or taking off honey. Frail settlements are especially powerless. The most effective way to safeguard such settlements is to diminish entrance openings with wooden blocks, grass, or straw. Specific "looting screens" shield little settlements or cores from attack in queen-raising activities. Once looting has become pestilence, it isn't easy to stop. A beekeeper can mainly assemble colonies back and decrease their passages and those of neighboring states, allowing the way of behaving to run its course. While entering an apiary, consistently check out the denying potential by keeping a steady consciousness of the bees' fervor level. If looting goes crazy, the two bees and beekeepers endure. Controlling this damaging behavior is one of the quintessential demonstrations of good beekeeping. Requeen excessively guarded states. Give a pound or two of honey every year to your neighbors. Diminishing the Preventiveness of various beekeeping procedures might decrease the protectiveness of a state.

Some include areas; others require a more profound comprehension of honeybee needs. Here are some primary techniques: Put only a few states in one area. Find hives in the direct sun. Generally, use smoke while controlling bees. Keep the region around the hive liberated from presented honey to limit burglarizing. Try not to move hives. Don't open hives during terrible weather conditions. Think about the Creatures. In country regions, guarantee that animals are not found nearby (some will generally involve beehives as scratching posts). Ponies are a specific issue. Their smell appears to insult bees, and while stinging beginnings, ponies frequently respond savagely and can harm themselves. Likewise, ponies are not as open-minded to toxins as people and can be killed without much of a stretch.

Ponies are the most important animals, and a stinging occurrence could cause a significant legitimate gamble. Wild creatures who could examine beehives incorporate bears and skunks.

Weather Conditions Issues

 Consider the climatic components in your district as you pick your site. In desert conditions, shade and windbreaks are fundamental, but not so in timberlands, where an excess of shade can be an issue and could bring about high mugginess in hives. There is proof that bees in the direct sun are not quite as guarded as those in concealed. Blustery destinations can be trying for bees. They don't fly well in blasts north of 15 miles (24.0 km) each hour and can become bewildered. Solid blasts might harm or bring down tall hives. Water accessibility is critical, as currently portrayed. Scrounge Accessibility Bees can't make honey or get by without legitimate rummage.

Luckily, the bugs are cosmopolitan and can exploit a great many plants. Numerous metropolitan regions support honeybees, yet the beekeeper frequently needs to learn about what the bugs may search on. The best wellspring of data on this is the nearby beekeepers' affiliation. Most areas can uphold a couple of colonys, yet pressures on the neighborhood plant assets increment as the quantity of hives increases. It is the neighborhood of a 1-to 2-mile (1.6 to 3.2 km) range of a settlement that matters. Areas with sufficient scrounge are challenging to track down and keep, and beekeepers will more often than not monitor them desirously. Deciding if a site will be great on a normal premise requires quite a while. One year's guard yield can be trailed by 1 or 2 years of lemon and the other way around. There are many plants that honeybees could visit, yet a couple is viewed as significant nectar sources in a particular locale. Many books have been distributed on major and minor honey plants and their reaches (see Assets). Rummage accessibility changes with the times. There are numerous instances of beekeeping activities that have needed to move in light of huge scope search misfortune. For instance, after The Second Great War, beekeepers needed to change areas from Alabama to Florida when their customary nectar source, the tulip poplar (Liriodendron tulipifera), was chopped down to help the conflict exertion. Then again, cotton honey has, as of late, prospered across the South, given the progress of the boll weevil destruction program. Before this happened, finding honeybees in the cotton nation was capital punishment due to the broad utilization of pesticides.

Saving Honeybee Scrounge

 Beekeepers ought to perceive that a couple of the many existing blossoming plants are significant nectar makers, albeit most will give, in any event, some nectar and dust to bees at specific seasons. Most of the significant honeybee plants are wild or wild. This makes beekeeping, such as fishing, an extractive industry. It generally depends on utilizing plants that are not developed and are dependent upon many impacts beyond the beekeeper's reach. In the Midwest, corn and soybean plantings have prompted the development of peripheral grounds, recently saved for most nectariferous plants.

Furthermore, in the South, endless suburbia and broad land waste for rural purposes keep diminishing areas occupied by plants valuable to honeybees. There needs to be a consistent settlement on the most proficient method to tackle diminishing nectar assets. Approaches include Enormous scope farming, planting of nectar-emitting crops Planting side of the road, recovered mining regions, and neglected land with nectariferous vegetation Cutting side of the road vegetation as opposed to treating with synthetics Creating plants that are prevalent nectar makers through hereditary designing, Presenting species that are demonstrated nectar makers in different locales. It will take participation among various gatherings engaged with keeping up with and restoring the wild plant scene to ration even an unobtrusive measure of honeybee scrounge from now on.

Chapter 5

Preparing to Build a Beehive

Now that Various Primers are far removed, now is the right time to think about some of the main choices fledgling beekeepers should make regarding beekeeping equipment. The contemplations in this part are no different for anybody keeping bees, yet may change relying upon the beekeeper's goals and experience. It is accessible all over the planet with Beekeeping equipment and supplies. In the US, a few stock houses have been open since the 1880s. New ones crop up sometimes. A few beekeepers make it a distraction to gather indexes and look at costs. The Web has now made many of these organizations effectively open through the Internet. One of the novice's difficulties is to avoid getting snatched up by every one of the choices. A decent rule is to remain with equipment that is steady with your goal and only to purchase a thing with grasping its worth and possible use. This book will adhere to the fundamentals, empowering the fledgling to begin the specialty in as basic a way as possible. Various other hive parts improve the standard hive body, including a base board, a honey super, a hive stand, a queen excluder, and a feeder. Likewise, all beekeepers need explicit devices in their weapons store, including a smoker, a hive device, and a defensive dress with a cloak to cover the face.

A kinder beekeeper Dr. Robert accomplished something few imagined at a new beekeepers' gathering. He had the option to hold the consideration of a crowd of people of committed Langstroth hive fans while portraying his encounters with the top bar hive (TBH). Right away, this sort of show fits those in formative apiculture who work with individuals with seriously restricted assets. The TBH has forever been viewed as a cut beneath the standard Langstroth hive innovation now set up in many regions, particularly where beekeeping is an enormous-scope business. Notwithstanding this, Dr. Mangum's comments were generally welcomed. They engaged the experimenter and hobbyist who lived inside each beekeeper. Dr. Mangum introduced his comments by saying that the TBH is all around intended for what he believes should do in beekeeping. The hive is cheap and can be made from scrap stumble, ideal for a striving youthful academician. It is likewise an adaptable system. Dr. Mangum portrayed lessening his bigger hives to diminish their weight with a buck saw — he just cut a few bars off the end. He then changed the slice end completely to a queen-raising core. This is maybe a definitive in hive adaptability. Envision taking a standard Langstroth 10-outline hive and changing it completely to an eight-outline model with a core left over in the deal! Even though Dr. Mangum does a great deal of queen raising, for which his TBHs are extraordinarily fit, he likewise stacks these hives toward the rear of his pickup truck and takes them, making a course for fertilization contracts. That's what he fights; regardless of his feelings of trepidation going against the norm, they commonly experience little harm during transport. TBH, beekeeping is simpler for the two bees and beekeepers; here are a few reasons: The brood is most positioned toward the front-entrance end of the hive, and the honeys situated in the back. Looking at the brood or taking off honeys, like this, less distressing on the bugs since one doesn't need to destroy the entire settlement. The top bars run into one another. They serve as a cover, decreasing material prerequisites and moderating weight. An external front of tin or cardboard is important, be that as it may, to safeguard the colony from dampness. Just the piece of the hive being worked is uncovered during control, which diminishes general protectiveness. At last, every one of Dr.Robert's hives is mounted on remains at the midsection level, holding him back from having to ceaselessly twist around. Top bar beekeeping isn't a great fit for everybody, except it merits a pursuit by anybody considering a kinder, gentler method for keeping bees.

The portable edge beehive is the groundwork of present-day beekeeping. This must be considered. Most oversaw beehives all over the planet. The aspects he utilized, a case obliging 10 edges, inches (24.3 cm) high (frequently called full profundity), have become the norm in the US The utilization of "standard" equipment, consequently, is the foundation of present-day beekeeping in the US; as a result, most equipment that anyone could hope to find to the beginner in this nation is compatible, reasonable, and promptly accessible. Beekeepers not utilizing standard or Langstroth equipment risk future issues when extending or consolidating their activities with others. Offered that guidance, beekeepers ought to realize that different systems are accessible and seem OK in certain circumstances. These incorporate hives because of eight casings, which consider simpler lifting for the administrator, and the shallower (6-inch [16.8 cm]) Dadant square box, for those wishing to widely exchange equipment. Another system acquiring disciples is the top bar hive, a profoundly unique plan utilized by certain beekeepers who favor a limited scale to deal with the specialty.

Edges and Establishment

Ten brushes fit in the standard hive body (brood chamber), each encompassed by a casing. The casing is planned with the goal that a basic hole, or honeybee space, of about K of an inch (0.97 cm) is left on all sides between it and the inside mass of the hive. The bees don't incorporate into or stick up this space. Edges can be wooden or plastic. The last option is becoming more OK to beekeepers, especially those worried about beeswax defilement from vermin medicines: plastic casings are presently promptly accessible, some with incorporated plastic establishment. Different blends are plastic edges with the wax establishment and wooden casings with the plastic establishment. Each edge has a top bar, a base bar, and two sidebars collected into a square shape that holds the brush. Outlines are fitted with sheets of beeswax establishment; appropriately named, it is the format from which bees assemble their home. The establishment is economically created out of beeswax or plastic and decorated with an example of hexagonal cells. The bees then, at that point, draw out the walls of the cells in light of the format (establishment). You can support the establishment with implanted wires or holding pins. For the most part, the embellished cells on the establishment measure 5.1 millimeters wide; in any case, a few beekeepers like to involve more modest establishments in the 4.9-millimeter range.

Building a Frame with Frame Spacers

Regardless of the many choices, a beekeeping novice should start by gathering wooden casings, embedding wax establishment, and wiring it for solidness.

Beeswax establishments embedded in wedge-style top bars and split base bars are the norm. 1. Eliminate the wedge from the top bar

2. Collect casings with paste and nails: each has a top bar, two sidebars with holding pin openings, and a base bar.

3. Embed prewired establishment (displayed here with inserted vertical wires with snares). Guarantee snares are ready from where the wedge was taken. 4. Supplant the wedge and nail it to the top bar against the wire snares.

About Base Establishment of Hive

The establishment is suitably named. It is designed beeswax given to honeybees to direct them as they construct brush. It likewise permits the bugs to save energy since it gives a reasonable setup of wax to make up the inward midrib of the brush. Various sorts of the establishment are sold, contingent upon the beekeeper's need. That utilized in the brood home is thicker than that sold stringently to create brush and honey. Bees favor their establishment to be beeswax, yet lately, plastic (covered and not covered with beeswax) has increased with some achievement. The establishment might be built up with vertical wires, which can deliver serious areas of strength for very valuable in outfits that concentrate honey at high velocities. An ongoing discussion about establishment centers around the cell size that outcomes from the layout gave. In nature, the hexagonal cells of the brush can change, going from 5.2 mm to 4.9 mm. This outcome in bees of various sizes might influence the state's conduct in some ways not completely perceived. Laborer honeybees were believed to be more useful and sound if created in bigger cells. Some accept this has fostered a honeybee that is too enormous. Enter the people who think the opportunity has arrived to decrease cell size in light of reports that Varroa less impacts more modest bees from the more modest establishment. Until now, there's very little direct logical proof of this, but rather the "little cell" people may be on to something. The reality of the situation will come out eventually. Base Board The base board, the floor of the hive, has various capabilities. Those accessible in supply houses are typically reversible, giving the choice of a little entry for winter and a bigger one until the end of the year. The base board is likewise basic for moving colonies. You can adjust the hive floor to fit different embellishments, for example, feeders, dust snares (for scratching dust off bees' legs as they enter), or dead-honeybee traps (to quantify settlement mortality). The baseboard has recently turned into a device for checking and controlling the Varroa bug. A lubed paper or "tacky board" put on the baseboard will gather parasites who tumble from the brood home. You can then count the parasites to decide the invasion level. The "open" or "screened" baseboard consider vermin that tumble off one or the other brood or grown-up bees to be forever taken out from the hive because once on the ground, they can't get back in.

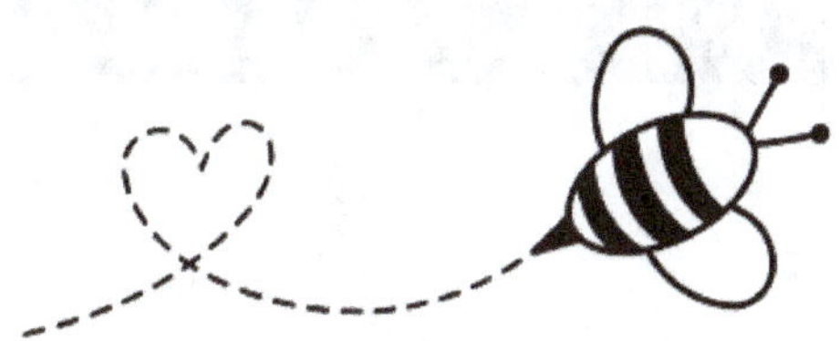

The queen excluder is a progression of equal bars dispersed so working drones can crush between them, yet the bigger queen can't. Generally, the beekeeper puts the excluder over the brood chamber to keep the queen from entering the honey super and laying eggs. This holds honey supers back from being defiled with brood since most beekeepers like to collect strong brushes of honey. Numerous beekeepers, in any case, consider this gadget a "honey excluder," accepting that workers with enormous nectar loads are likewise obstructed in their endeavors to move uninhibitedly around the hive, consequently diminishing the last honey creation. At times, this gives off an impression of being valid, particularly with a feeble nectar stream intensified by a little laborer populace. Novices, in this way, ought to utilize it sparingly and when solid nectar streams with ideal honeybee populace warrant its utilization.

Scarcely any inquiries in beekeeping circles appear to incite as much contention as whether to utilize queen excluders. Vocal advocates exist on the two sides of the issue. A predominant thought by the individuals who denounce the utilization of excluders is that engorged bees are greater and thus will only fit through the painstakingly estimated supports of the excluder. The rationale follows that a reasonable setup of Honey is kept from entering the super. The board strategies by these beekeepers keep away from issues that queen excluders are intended to dispense with; for example, the presence of one or the other brood or queen in honey supers eliminated for extraction. Once a maturing nectar is kept over the brood home, the queen seldom crosses it to lay eggs. Beekeepers who would rather not risk losing the queen or battling with brood in, to some extent, occupied supers at extraction time, in any case, stick rigorously to excluders. Also, they contend that excluders don't unfavorably influence honey creation. Keeping the queen to a specific region of the home is likewise standard beekeeping practice in many queen-raising activities. This would be unthinkable without excluder innovation. Who's thinking correctly about excluder use?

The two sides are. Whether to involve innovation relies upon its apparent utility in unambiguous honeys. Covers Most stockpile houses sell extending external and ventilated internal hive covers. Both are standard items the beginner ought to obtain. Some enormous-scope beekeepers select what is known as a transitory cover, frequently just a piece of pressed wood utilized without an internal cover. These are less expensive than the extending assortment, yet all at once, not as secure. A benefit, notwithstanding, is that they permit hives to be stacked together firmly when moved by truck. The inward cover can be utilized in various ways. Most have an edge and an elliptical opening in the middle to oblige a honeybee to escape a gadget. The edge is indented, offering the chance for expanded ventilation.

In many regions, the hive stand is fundamental. It keeps wooden hives off the ground, shielding them from clamminess, dry decay, termites, and different bugs. Substantial blocks, elastic tires, and synthetically saved wooden rails could be generally used to develop a stand. In regions where subterranean insects are an issue, the hive stand can be a rack roosted on slender legs that are embedded into jars of oil or water, which keep these bugs under control.

The Hive Scale

The hive scale is a valuable device for the fledgling beekeeper to decide how a settlement manages without upsetting the bees. A hive that appears light in the center of a significant nectar stream, or when hives around it are observably heavier, is a prompt reason to worry. Stage scales are customarily used to weigh hives. Sharp beekeepers, in any case, have concocted various sorts of gadgets in light of restrooms and different scales and have portrayed and shown them in writing. In this period of P.C.s, a few beekeepers interface scales straightforwardly to electronic screens. Gauging a hive is worthless, in any case, if the obligation to record the information isn't there. Using the Web, up to one public drive to gather scale hive information has been carried out. Feeders The Boardman feeder, a modified glass container that squeezes into the entry of the hive, is generally regularly offered to fledgling beekeepers, and it is a decent decision for novices. A significant benefit of the Boardman is that the syrup level can be checked and the supply recharged without destroying the hive. There are a few burdens to this feeder, nonetheless. It doesn't hold much food and might instigate ransacking conduct when utilized with frail states. Different taking-care plans incorporate adjusted outlines (division board feeders) and top feeders, some of which can be very intricate. Plastic sacks loaded with syrup can be penetrated with little openings and altered over the edges.

The Smoker

The smoker is a significant and fundamental beekeeping instrument, even though you might discover a few beekeepers who guarantee not to require it. Smoke stifles a protective way of behaving, and you ought to have a smoker lit any time you are in the bee yard. Smokers generally come in two sizes, 4×7 inches (10.2×17.8 cm) and 4×10 inches (10.2×25.4 cm). Tempered steel models will endure longer. A safeguard to forestall direct contact of the hot barrel with the beekeeper's skin and garments is generally attractive. A few shapes have been conceived, and you can paint yours in splendid varieties, so it may be effortlessly seen when you abandon it in the field. When necessary, different things like screwdrivers can be utilized to control states, yet they can harm your equipment. Novices should purchase two apparatuses, one for each hive they begin with, as suggested in this book.

Beekeeper Attire

There is an extensive variety of defensive dress accessible to beekeepers. It is generally tricky to wear more while starting a hive control since taking off is typically simpler than putting on beekeeper clothing. Contingent upon your responses to stings, you might choose to be completely secured. Just a single thing is fundamental, in any case: you can eliminate any piece while examining bees, except the shroud. Even though you can recuperate from the impacts of stings regardless of where they could happen on the body, is it a good idea for you to get stung on the eyeball? Serious harm would be finished, bringing about visual impairment. Never control bees without a shroud.

Shroud and Cap

An assortment of headgear is offered to fledgling beekeepers to help that basic frill, the cover. Wear a cloak consistently to shield your face and head from stings where they are generally difficult. Shrouds are intended to be worn with protective caps and caps or, at times, without a cap (the Alexander model). The square collapsing type is more sturdy over the long haul since you can undoubtedly stash it when you are not utilizing it.

Chapter 6

Introducing Bees in Colony

Bundle Bees

Bundle bees are delivered in the southern states and California for delivery to northern beekeepers who wish to reinforce frail colonies or lay out new colonies in the spring. Bundles are accessible in 2-, 3-, or 5-pound sizes. The most famous bundles are the 2-and 3-pound sizes. Each pound addresses around 3,500 bees. A recently mated queen is incorporated to be utilized for growing new colonies. Bundles expected for reinforcing powerless colonies might be requested regardless of a queen.

You ought to arrange bundles in January or February to guarantee ideal conveyance in late-winter (April). Assuming that you are introducing bundles on drawn combs containing honey and dust, you can do as such toward the beginning of April; in the event that you are introducing them on comb establishment, you ought to have them show up later than expected April or early May. Beekeepers in northern regions might wish to defer shipments for a long time. Bundle bees could bite the dust whenever introduced on establishment in temperatures underneath 57°F (14°C) on the grounds that too couple of bees will actually want to break bunch and move to syrup feeders. Bees grouped on combs of honey, then again, don't need to break bunch to eat. Standard wooden transportation confines measure around 6 x 10 x 16 creeps with wire screen on the long sides for ventilation. A can containing a food supply of 50% sugar syrup is situated in the enclosure. A couple of little openings in the lower part of the can permit the bees to pull out the syrup. A youthful mated queen is housed in a different enclosure that is suspended at the highest point of the bundle close to the feeder can. A few working drones (orderlies) regularly are confined with the queen to really focus on her. Queen confines as a rule are provided with a food wellspring of sugar sweets. A bug treatment strip may likewise be suspended from the highest point of the enclosure or stapled to the rear of the queen confine. Bundles are propped a few inches separated to shield them from swarming and overheating during shipment. You can get bundles from a nearby beekeeper or supply vendor who has bought them in mass from a trustworthy bundle maker, or you can arrange them straightforwardly from the maker and have them transported by U.S. mail. You ought to caution authorities at the mail center about the normal date of appearance and ought to demand quick warning. Upon appearance, you ought to examine bundle bees for strange quantities of dead bees. Some honey bee mortality is typical, yet when dead bees gather in excess of a 1 / 2 inch in the lower part of the transportation enclosure or when queens are washed up, you ought to record a harm guarantee with the postal representative right away, taking note of the state of the bundle. You ought to then send this assertion to the bundle maker so misfortunes might be supplanted.

Quite possibly of the main thought in fostering major areas of strength for a from a bundle is to supply a lot of food until a solid nectar stream starts. Except if you introduce your bundles on drawn combs containing adequate honey and dust (taken from existing colonies or from stockpiling), you ought to want to take care of the bees promptly upon establishment and keep taking care of them until they can fight for themselves. This is basically significant when introduce ing bundles on establishment. There are a few proficient approaches to taking care of sugar syrup to your colonies. One of the least demanding strategies for getting food to colonies hived from bundles is to transform a feeder can or plastic container over the opening in the inward cover. You can make this feeder by punching ten to fifteen little nail openings in the top of a container or can with removable top, (for example, espresso can or clean paint can). Try not to leave the hive top and feeder can uncovered — put an unfilled hive body on top of the hive body with the bees to encase the feeder and supplant the external cover. You can likewise introduce bundle bees in a hive body over a twofold screen put on top of a laid areas of strength for out. The glow of the laid out colony on the base works on the advancement of the new colony. Hive the bundle by shaking the bees from the steel trailer and direct delivery the queen (as portrayed already). You want to give an entry to the back of the hive for the new colony and feed it sugar syrup as you would different hives laid out from bundles.

Gathering Swarms

Gathering honey bee swarms in the spring is an astounding method for supplanting winter misfortunes, reinforce frail colonies, or begin new ones. Essential multitudes are significant; they might contain upwards of 25,000 bees in addition to the queen. In correlation, a 3-pound bundle will number roughly 10,500 bees. Three contemplations to remember before endeavoring to gather a multitude are (1) how long the multitude has been there, (2) where the multitude is found, and (3) its size. Swarms typically group on a tree appendage.

The two amateurs and laid out beekeepers ought to choose every apiary site cautiously. All through the scrounging season, nectar and dust sources should be inside a brief distance (about 1 mile) of the hives. Dust is fundamental for brood raising, and nectar (honey) is the bees' essential wellspring of energy. While bees can be kept practically anyplace, enormous groupings of flower sources (and crowded colonies) are expected to deliver huge honey yields. Bees likewise need a wellspring of new water so they can weaken honey, manage hive temperature, condense solidified honey, and raise brood. In the event that a water supply isn't accessible inside 1/4 mile of the hives, you can give a tank or container of water with a drifting board or squashed rock for the bees to arrive on. The water source needn't bother with to be "unadulterated." Bees are less bad tempered and simpler to deal with when situated in the open where they can get a lot of daylight. Conceal from trees impedes the trip of workers and frustrates tracking down the queen and seeing eggs inside the cells. A southern or easterly openness gives colonies greatest daylight over the course of the day. The apiary is best arranged close to normal breeze security like slopes, structures, or evergreens (Figure 23). Different necessities are dry ground and great air seepage. Keep away from blustery, uncovered ridges or destinations close to the bank of a stream that could possibly flood. You ought to likewise keep away from apiary areas in vigorously concealed woods or in a sodden base land since overabundance dampness and less daylight impede the trip of the bees and energize improvement of such honey bee diseases as nosema and EFB. Your openness to the apiary is significant — maybe the main figure apiary area since you should visit it all through the year in a wide range of climate.

Chapter 7

Honey Production and Handling

Types of Honey

Honeys showcased in five essential structures: segment comb, cut-comb, lump, finely solidified or creamed, and extricated (fluid) Honey. Equipment necessities and the board change with the kind of Honey you intend to create. Most beekeepers produce removed honey. More overflow honey can be gotten from statesman matured for extricated Honey than those oversaw for comb honey. Combs utilized for removed honey creation require support like wires, strings, or plastic sheets. These materials, in any case, wouldn't be tasteful to the purchaser of piece, segment, or cut-comb Honey. The development of separated Honey requires exceptional equipment for uncapping combs and eliminating Honey from cells. Cut-comb and piece honey creation require comparative administration, and the greater part of the equipment can be utilized conversely. Cut-comb Honey, the most affordable to deliver, is great for the fledgling. Area comb honey requires specific equipment, extraordinary administration, and a plentiful nectar stream for good returns and isn't regularly suggested for novices.

Area Comb Honey

Segment comb honeys created and sold in the comb in either little wooden (41/4 x 41/4 inches or 4 x 5 inches) or round plastic areas. Comb honey makers practice two fundamental systems of spring the executives, contingent upon the size of brood chambers utilized. These two systems have various administration choices. The most straightforward strategy for creating comb honeys to winter the bees in an 11/2 - story brood chamber. A subsequent shallow box might be added about the time maples or natural product trees are in sprout if the spring stream is weighty. At the point when the principal nectar stream has begun, place the queen in the full-profundity hive body and supplement a queen excluder underneath the shallow super. When the bees have put away Honey in everything except the two external edges of the shallow super, the colony is prepared for a segment super. Place the part beneath the queen excluder. Queens rarely lay eggs in the areas since the space is partitioned into little compartments. An excluder set under a segment, at times, hinders crafted by the bees in the segments, yet it doesn't appear to impede work in supers, particularly on the off chance that the bees are, as of now, working in them. The Honey and brood in the super shallow urge the bees to work in part rapidly since they are now familiar with going into the abovementioned. Add an extra area when the bees have started working in the segments' external lines in the last superadded. Place void supers under the excluder and on top of the other area supers. If colonies are solid and the nectar stream is great, a colony might fill upwards of four or five-segment supers. Putting void area supers over those somewhat filled supers keeps the workers more coordinated. Workers do a complete occupation of filling and fixing the segments as they work up into the unfilled super. On the other hand, you might raise the principal super when it is half fixed and places a, to some degree, drawn under. This control of supers lessens the "travel stain" gamble on the part capping from the brood chamber underneath. At the point when supers are added excessively fast, they might contain somewhat filled segments with almost no market esteem. Capable creation of comb honey takes practice and great nectar streams. The second administration system for comb honey creation includes permitting a colony to completely venture into two full-profundity brood chambers. Beekeepers with extensive experience can effectively utilize this methodology. The twofold brood chamber gives the queen a wealth of room for egg-laying and the Colony a lot of room for Honey Extra space for brood raising is, by and large, not required. In late April, the brood chambers are typically switched. The Colony takes care of light sugar syrup to invigorate colony development, as portrayed in the section on spring, the executives. Except if there is an extremely weighty nectar stream, bees won't fill segment combs effectively when the part supers are set over two hive bodies; they are bound to fill the two brood chambers and afterward swarm. Hence, one brood chamber is typically removed, and the bees swarmed when the vital Honey stream starts. You can decide the start of the stream by the presence of new white wax on the combs and the top bars of the brood outlines. New nectar that drops out effectively when shaken will likewise be in the brood combs. Place the queen in the lower hive body with five edges of fixed brood and five void combs when you lessen two-story colonies to a solitary story. Place the excess combs in a vacant hive body after shaking around 66% of the bees from them before the lower hive body. Give the Colony a couple of segment supers, contingent upon its solidarity. You might set the subsequent hive body on a frail colony or use it to foster another colony. For best outcomes, snare segment supers with a couple of halfway-drawn areas saved from the past season. While utilizing lure segments, place four of them in the middle line of areas. Lure segments utilized in the underlying super are often poor in quality when filled and shouldn't go to advertise. At the point when bees are all around, begin in the peripheral segments of the super, and there is motivation to expect the nectar stream to proceed, add another super. Try not to give room before the bees need it. Add more supers reasonably around the finish of the nectar stream; if not, many areas will be begun that won't be done in attractive condition. Place the vacant really over the somewhat filled one as portrayed already. Bees will then complete the segments as they climb. At the point when a segment super is fixed through and through, except the two external columns, eliminate it and spot it over a honeybee escape. Try not to eliminate areas with smoke because the bees will bite minuscule openings through the cappings during smoking, creating flawed segments. Furthermore, the Honey might assimilate the kind of smoke and become disagreeable. Utilize incomplete segments from the external lines as snare segments in the following super, when they are given to the bees. When the stream is about finished, eliminate the segments and store them in a cooler for the following season. To decide when a segment super is full, take a gander at the bottoms of the segments since these are the last parts to be fixed. Adding a very like clockwork is prudent if nectar is coming in quickly. Adding supers at a legitimate time requires great judgment and information on neighborhood conditions. Similarly as significant as adding supers at the fortunate time is their evacuation in an ideal design. If a completed super is left on the hive too long after the combs are fixed, cappings will become messy or "travel-stained." Five to eight days are expected to mature nectar; in this way, all Honey ought to be ready 12 days after the stream stops. Inspect the brood region of all colonies delivering comb honey for queen cells each 8 to 10 days during the amassing season. This is particularly significant when bees from two brood boxes have been decreased to a solitary brood box. Obliterate all queen cells; the Colony might crowd if any cells are missed. Segment boxes for creating comb honey ought to be collapsed, loaded up with the establishment, and set in supers just not long from now before use. Assuming they are made excessively far ahead, they might be less appealing to the bees or can twist. The crates will then, at that point, be prepared for guaranteed use during the nectar stream. While collapsing the wooden segment boxes, saturate all depressions with a clammy wipe, cloth, or fine shower of water or steam to forestall breakage at the corners. Ensure the sheets of slender excess establishments are similar in length and profundity to the four segments and situated in the crates. Divide wooden areas and plastic segments to gather more quicker than strong segment boxes. Place four collapsed segments in the part holder, with the split side up. Spread the highest points of every one of the four areas. Drop the sheet of establishment into the space left between the two parts of the four segments and push the areas together to hold the establishment tight. Gadgets for placing establishments in split segments are accessible from honeybee supply sellers. Comb honeys more appealing and monetarily created when you utilize full sheets of the establishment. Assuming you lean toward strong areas, cut the establishment around 1/ 8 inch more modestly every way than within the segments. Plan four or eight blocks, each block 1/8 inch more modest than within the part. Blocks ought to be 1/8 inch more slender than 1/2 the profundity of the part (e.g., blocks for 17/8 - inch segments ought to be 7/8 inch thick). Place the areas over the blocks and lay one sheet of establishment on each block. Then, at that point, slide a hot cutting edge between the part and the foundation. Push the establishment against the hot edge as you pull out the cutting edge. The dissolved edge of the establishment will slide against the segment and stick as it cools. The cutting edge might be a wide scratching blade or comparative device, but it should be wider than the establishment. Embed the segments into a part holder. The establishment ought to then hang liberated from the two sides and exceptionally close to the focal point of the segment. It should draw near around 1/8 to 1/4 inch of the base. The establishment shouldn't lay on the base since it extends a little in the hive. If it arrives at the base, it might clasp and twist the subsequent comb. Try not to put the segments in supers until they are required. A couple of ineffectively secured sheets of establishment tumbling down after it has been put away or given to the bees might cause extensive misfortune and burden after the bees have filled the super. Paint the highest points of parted or strong areas with liquefied paraffin or cover them with veiling tape so the bees can't stain the segments. Dissolve the wax in a twofold evaporator with water in the base. Never put a dish of paraffin wax on a burner because the wax might burst into flames or get excessively hot. Assuming the wax is excessively hot, it obscures the segments. Eliminate completed supers cautiously and without smoke to avoid harming the cappings. Eliminate wedges or materials used to hold areas together in the super. Put them on a 1-inch-thick board or block of wood sufficiently huge to slip inside the super. Push down on the hive body to drive the segment holders and separators out of the super body so they can be promptly isolated and the areas scratched and cleaned. Cautiously scratch areas free of propolis and eliminate the concealing tape. This activity requires a lot of care to try not to harm attractive segments. In the wake of cleaning, gauging, and evaluating segments, encloses them in cellophane sacks or cardboard containers with a cellophane front. The cardboard container is leaned toward it because it better safeguards the fragile area. Customary wooden segments are challenging to create in amount or quality. A few beekeepers who need to deliver segment honey have picked Ross Rounds Segment Comb (now and then Round Segment Comb). When filled, you should buy an extraordinary very with round plastic rings holding 12 ounces of Honey The part super will hold 36 of these round segments (versus 24 in the wooden area super), and there is no scratching, and assembly of establishment is moderately simple. Alluring plastic covers and a specialty name offer deals request. The item sells for about a similar cost as the square segments, yet because of more modest sizes, more are delivered and of better quality; there doesn't appear to be shopper hesitance to buy the plastic areas when accessible. A half-comb tape for delivering comb honey is likewise accessible. It is one side of the typically two-sided comb held in plastic boxes in specialty supers. Bundling is appealing, and there is no gathering required. A super delivers 40 units when filled. Market acknowledgment has yet to be tried; selling great as a specialty product is normal. Comb honey ought not to be delivered in that frame of mind from the nectar of wild blossoms since this Honey solidifies more rapidly than most summer honey.

Lump Honey Piece honeys regularly created in shallow casing supers with the slim excess establishment in the edges, as you would deliver cut-comb

Honey wooden wedge or dot of softened beeswax holds the establishment at the highest point of the casings. Bees are overseen similarly as expected for comb honey; supers are added over the queen excluder. Permit the queen to extend her brood in a shallow or medium hive over the full-profundity hive body in late winter or utilize two standard profound hive bodies. When the upper hive body is around 3/4 loaded with Honey and brood, you should drop the queen down to the lower standard hive body and restrict her there with a queen excluder. Following 4 or 5 days, place a vacant super with edges and establishment underneath the full upper one over the queen excluder so the queen can't lay eggs in it. Add more supers depending on the situation. Put an unfilled one on top of the stack until the last one is 3/4 full, and then invert them. Keep adding supers along these lines. Eliminate supers utilizing a honeybee getaway and control amassing; similarly, you would control colonies overseen for comb honey. After eliminating covered outlines, cut the comb into enormous pieces to fill a wide-mouth container. Prior to placing them in the containers, put the pieces on a screen and let them channel over a channel skillet in a warm space for a few hours. Occupy void spaces in the container around the comb with fluid Honey that has been warmed to 140°F (60°C) before pouring it over the comb in the container. Warming the fluid, Honey defers crystallization for a long time. Sorts of Honey that take shape rapidly, like hay or wild aster, are not appropriate to piece honey creation since they solidify on the food merchant's racks. Mark the item available to be purchased as lump honey.

Honey Evacuation and Handling

Eliminating Honey from the Colony While eliminating the honey yield throughout the mid-year, make certain to leave sufficient stores for the bees in the event of an absence of a fall crop. A decent rule is consistently leaving brimming with honey with the bees. Eliminating the completely covered supers before the honey stream has stopped is more averse to starting looting conduct. Extraordinary looting might happen if you keep eliminating the supers until after the stream is all finished. Likewise, the expulsion of the spring and summer honey yield not long before the beginning of the goldenrod stream will permit you to keep the honey isolated by flavor. Generally, summer honey is lighter and milder in flavor than the more obscure, more extravagant seasoned fall kinds of honey. Fall kinds of Honey frequently take shape quickly, which could make a few issues at extraction time if you stand by to eliminate the whole yield on the double. The fall honey yield should not be eliminated until after a block of killing ice. In a perfect world, edges and supers should be eliminated when completely covered. However, they should be something like 3/4 covered before removal. Attempt to keep the gather of to some extent covered edges to a base; any other way you might generally disapprove of high-dampness Honey You can utilize a few distinct methods to eliminate the supers from the colonies, contingent upon the size of your activity. Use smoke sparingly while eliminating combs or potential supers in light of its impact on the kind of honey. While collecting a couple of combs or supers honey, shaking and brushing the bees from the combs might be the most viable technique. To delicately do this, open the top and smoke the bees. Eliminate each edge in turn and, holding each firmly by the finishes of the top bar, give it at least one fast shake descending in the air over the open Colony or inside an open space of the super to eliminate the vast majority of the bees. The excess bees can be dismissed the comb with a brush or a cluster of grass before the hive. Place the collected outlines in covered supers to abstain from looting. An option cheap, low-tech strategy for eliminating honey supers is to utilize a honeybee escape. Utilizing Honey bee escapes requires two excursions to the apiary — one to put on the departures and the other to eliminate the honey. Two unique departures are ordinarily utilized today. The Doorman honeybee escape is a metal or plastic gadget set worse than broke of the internal cover. The internal cover is then positioned between the eliminated supers and those left on the bees. The three-sided honeybee escape is the size of an internal cover fitted with two arrangements of triangles that go about as a labyrinth. It is set similarly situated as portrayed previously. In the two cases, the departures are one-way entryways, permitting bees to drop down yet not back up into the supers once they are beneath the break. Since they permit just a single honeybee at once through the break, Watchman honeybee escapes are slow and can sometimes become stopped. This isn't an issue with three-sided getaways; be that as it may, they are more expensive. The getaways are normally left on the colonies for 2 or 3 days to allow the bees to clear the supers. Cool night temperatures are important to allure the bees to pass on the supers to join the hotter brood underneath. If the supers are not honeybee tight, the Honey over the getaway sheets will probably be burglarized by different colonies. Assuming that the stickiness is high when the breaks are on the colonies, the Honey might get some dampness. When even a limited quantity of brood is available in the supers, bees will stay with the brood, and the breaks won't work, so the leftover bees should be shaken or gotten over. Honeybee gets away from function admirably whenever utilized with queen excluders. The excluder keeps the queen from laying in the honey supers, and when the time has come to eliminate the Honey, a honeybee escape (and internal cover if utilizing a Doorman honeybee escape) is filled in for the excluder. The most straightforward opportunity to eliminate supers of Honey got free from bees with a honeybee escape is in the early morning before the bees are flying. A third option is to utilize a substance, honeybee repellent and smoke board. Synthetic substances utilized for eliminating Honey include Benz aldehyde (oil of almond), butyric anhydride, and an oil/spice combination. Sprinkle a couple of drops of the compound on a smoke board, which is made by extending a weighty piece of material over an edge that is the size of the inward cover. Cover the top with a piece of sheet metal to support it and paint it dark so it ingests heat from the sun. Place the smoke board over the full supers. The exhaust drives the bees descending. Blowing a couple of puffs of smoke over the top bars before adding the smoke board will begin the bees descending, making them less inclined to be confounded. The board ought to stay on the main long enough to get the bees out, typically 3 to 5 minutes under ideal circumstances.

Various variables impact the nature of Honey from when it is taken out from the Colony until it is sold for human utilization. Whether the activity is little or enormous, delivering a last-pack matured item that is of the best and appealing to the shopper is significant. Honeys considered top quality soon after the bees fix it in the comb. Legitimate taking care of during extraction and handling can deliver fluid honey with just a slight loss of value. While certain shoppers need crude or natural honey(fluid honey that has not been separated or warmed), most market outlets require honey with a long timeframe of realistic usability. Accordingly, stressing and some warming are prudent to postpone granulation and forestall aging. The result ought to be all around stressed, low in moisture, liberated from unfamiliar flavors and pollutants, and ought to hold its unique fragile flavor and fragrance. Different variables that bring down the nature of the result are the overabundance of air pockets, dust, and pieces of wax integrated into the honey during extraction. Honey quality is impacted most by warming and dampness content. Never during handling should the honeybee overheated. An overabundance of heat chemically separates the laevulose sugar, obscuring the Honey and disposing of the normal, unpredictable flavors that make Honey extraordinary. Honeys hygroscopic — it promptly assimilates dampness from clammy air and loses it to dry air. Dampness even goes through the wax cappings. In this way, the level of readiness at the time the Honey is eliminated from the Colony is generally connected with the predominant barometrical stickiness.

Ingestion of dampness reduces the grade and timeframe of realistic usability of the Honey High-dampness Honey might age. After removing honey supers from the Colony, they should be held in a warm, dry region until extraction. The best opportunity to eliminate abundance dampness from Honey, assuming essential, is while the Honey is still in the comb. Store the supers in a warm room at 75°-80°F (23-26°C) for several days or stack them over a light so the intensity misses through the casings and warms the Honey Safeguard the light so that Honey and wax won't trickle straightforwardly onto the bulb. An electric fan can be utilized to course the air in the room.

On the other hand, you can utilize a perfect vacuum cleaner to compel air straightforwardly through a heap of supers; cut an opening in a sufficiently enormous to allow the section of the vacuum hose. Over this super, stack seven or eight supers of Honey and turn on the vacuum with the goal that it will compel an enormous volume of warm, dry air through the combs. How much dampness is eliminated will be connected with the general mugginess and volume of circling air. In enormous business activities, supers ordinarily are set in hot rooms before extraction. Warming Honeywell likewise accelerate the extraction cycle. Honey held for a couple of days at room temperatures between 80°F (26°C) and 90°F (32°C) is great for fast total extraction.

Extraction Procedures

Eliminating Honey from the combs is challenging for the specialist since there is no straightforward, slick, and inexpensive approach to doing as such. The crudest approach to eliminating Honey from the cells (destruct gathers) is to cut the combs from the casings and let the honey channel from the cells. To facilitate the cycle, put bits of honey-filled comb in a fine cross-section pack, then, at that point, pulverize the combs and press the Honey out the hard way. Then strain the last blend through a coarse sifter or fabric like cheesecloth. The best technique for creating fluid Honey(non-destruct reaps) requires an extractor that utilizes outward power to turn the Honey from the cells. Different sorts and sizes of honey extractors are produced economically. You might buy an extractor, lease the equipment, or find a beekeeper that does custom separating or construct an extractor. The most important phase in extraction is the evacuation of wax capping. Uncap the two sides of the honey-filled combs with an uncapping fork or sharp blade warmed by power, steam, or by dunking it in steaming hot water. Cut a slim layer of wax and Honey from the outer layer of each comb with a to-and-fro cutting development while you hold the blade against the top and base bars of the edge. First, uncap one side, turn the casing and uncap the opposite side.

Turn the end bar of the casing on the place of a nail upheld by a piece of wood lying across the highest point of the holder that gets the capping. Hold the casing at a point so the capping fall liberated from the comb into the compartment underneath. Utilize an uncapping fork (likewise named a capping scratcher) to tear open the capping in the low region of the comb not arrived at by the blade. Power unpapers with vibrating blades and programmed uncapping machines are accessible for enormous business activities. Since the capping contain a lot of Honey after they are cut from the combs, having some approach to isolating the Honey from the wax is significant. Permitting the capping to deplete into a screened box or wire bushel is a helpful choice for the little administrator. The most straightforward uncapping box for emptying the capping is made from a perfect hive body with a screen or queen excluder joined at the base or a metal/plastic holder fitted with a screen supplement to permit Honey pool underneath the capping suspended on the screen. This unit is put over a tank with the goal that the honeys gathered underneath. In bigger honeys, covering meters or spinners are utilized for recovering the Honey.

Limited-scope beekeepers frequently utilize a two-or four-outline container extractor, which might be either reversible or nonreversible. Place the uncapped combs upward in the bins that help them. In the nonreversible kind, you should turn around the combs by hand to disengage the Honey from the opposite side of the comb. Reversible extractors have containers that turn to disengage the initial one and then the opposite side of the comb without lifting and switching outlines. Either the hand-or power-driven extractors are turned gradually from the beginning. If the extractor is turned too quickly, the heaviness of the Honey might break the combs. The combs are turned until about 50% of the honeys removed from the main side.

Then, at that point, the combs are switched and turned until the subsequent side is separated. At long last, the combs are switched a subsequent time and the excess Honey from the underlying side is taken out. The time expected to toss Honey from the combs relies upon the thickness and temperature of the Honey Watch the side of the tank to see when the Honey prevents moving from the combs to decide when the extraction is finished. Huge outspread extractors from six to eighty edges are utilized in part-time and business activities. Combs don't need to be switched while utilizing a spiral extractor since Honeys tossed out of the two sides simultaneously. To accomplish this, the combs are organized in the extractors like spokes in a wheel, with the top bars outwardly.

Spiral extractors should be turned at a high pace, so very much-built outlines and secure combs are an unquestionable requirement. After the Honeys separated, it contains air pockets, dust, and pieces of wax. Avoid over-the-top dust by keeping brood combs out of the honey supers. Strain the Honey through a few layers of cheesecloth or a solitary layer of nylon in the wake of separating it from the combs. This technique eliminates most contaminations and sections of wax. Honeywell assimilate scents and flavors quickly if the materials that produce them are not stressed from the honey. Except if most of the wax has been taken out during extraction, the Honey's flavor might be hindered, particularly if, during pressing, the Honey is warmed past the dissolving point of the wax. After stressing, keep the Honey in a settling tank for 2 to 3 days to permit most of the air pockets and little unfamiliar particles to ascend to the top. The subsequent froth ought to then be skimmed off before packaging. Utilize a honey entryway at the lower part of the tank for filling containers or jars. In business honeys, the Honey regularly is siphoned through enormous breadth pipes with a lower outfitted honey siphon from the extractor(s) into a sump tank where a progression of bewilders takes out a large portion of the wax and pollutants. Bigger beekeepers sell or store their Honey in 5-gallon metal jars/plastic pails or drums holding 55 gallons.

Chapter 8

Storage of Honey

Capacity temperatures and the length of capacity can influence honey quality. Changes in handled honey are kept to a healthy level if the honeys put away at temperatures of 70°-75°F (21°-24°C). Natural honeys best put away below 50°F (10°C). Indeed, even at room temperature, honey slowly becomes hazier and changes flavor and creation. Contrasts will be noticeable in under one year. Both daylight and counterfeit light further influence honeypot away in clear glass bottles. Keep fluid honey in a cooler at 0°F (- 18°C) for long-haul stockpiling. Keep finely solidified or creamed honey in a fridge or in comparable cool conditions. Fridge temperatures prompt separated honey and honey in the comb to crush rapidly.

Liquefying Granulated Honey

Honey with coarse, abrasive precious stones is unwanted to the customer. To reliquary, place compartments of solidified honey in a hot, dry chamber or a hot water shower until all precious stones are condensed. If you utilize a dry chamber, the names on the containers won't be harmed. Try not to warm the honey higher than 145°F (63°C) because it sears without any problem. Buckwheat honey might consume at 140°F (60°C). Honey should be cooled when it turns out to be clear to prevent staining and loss of flavor. When warmed in 60-pound jars or pails kept straight up, you should mix the honey habitually; in any case, the honey close to beyond the can will be overheated before the middle is fluid. If you place the compartment in a hot water shower, the water ought to approach the highest point of the holder.

Bundling and Naming

Honeys stuffed and promoted in a wide assortment of compartments. Honeybee supply lists show compartments in many shapes and sizes reasonable for a wide range of business sectors. A typical retail bundle for fluid honeys the queen line-type glass container, accessible in various sizes from $1/2$ pounds as much to 4 pounds. The canning container is well known in certain areas, holding $1/2$ 16 ounces, half a quart, and a quart. One-piece covers are accessible for these containers. Various round containers will hold from 4 ounces to as much as 5 pounds of honey. Plastic containers are becoming more famous and accessible in many shapes and sizes, including the queen-line type. Curiosity holders incorporate hexagonal glass shapes, classical styles, plastic press cylinders, skeps, and the consistently famous honey bear. Honey, in bigger amounts, can be sold in plastic buckets and containers that hold from 3 pounds up to 2 gallons. The 5-gallon (60-pound) can is an extremely well-known method for bundling honey available to be purchased by the food business. One of the biggest holders is the 55-gallon drum with a food-grade lining. Honey to be pressed can be warmed to forestall maturation and granulation to expand its timeframe of realistic usability. Wash and air-dry all glass holders before filling. For ease in filling, the honey ought to be warm. You ought to hold the compartments at a point to let the honey once over the side to forestall the consolidation of air bubbles. If not, foam and air pockets will gather on the top surface of the honey, showing up. Fill glass containers over the globule that goes around the container. Fill all compartments so there is no hole between the cover's lower part and the honey's outer layer. Abstain from overloading. Wash off any honey that spills outwardly off the compartment or onto the strings. A tacky compartment is terrible enough for the client yet turns out to be even less alluring when covered with dust while sitting on the rack. All honey to be sold, as well as honey that will be parted with, ought to have an appealing name. A straightforward plan with at least two tones that complement the shade of the honeys alluring. A wide range of alluring marks is accessible from honeybee supply vendors (see supplement). Such marks can be engraved with your name and address for a marginally higher expense. All names need to observe the government mark regulations. State regulations, by and large, keep the government regulations. You can get a duplicate of both government and state regulations from your state Branch of Horticulture. It is vital to duplicate these regulations before planning a mark or, in any event, utilizing the pre-printed names accessible from honeybee supply vendors. Different exclusions exist for little or strangely molded compartments. As a rule, a name should have the accompanying: • "honey" should be truly noticeable — the biggest letters on the mark. You can name this source if you can show that a botanical source is the dominating one, generally finished with a dust examination. If not, "wildflower" would be the main assignment (however, the botanical source isn't needed). It is passable, for instance, to say "contains clover" or "contains basswood" on the off chance that you are sure of a potential flower source. • The net weight should be in the lower third of the front name in simple to-understand type. If the net weight is under a pound, it tends to be expressed in ounces and grams. Assuming that the net weight is 1 pound or over, it should be expressed in pounds, identical ounces, and grams. For instance: 16 oz (1 lb.) 454 g. For over 4 pounds, expressing ounces isn't required. However, grams or kilograms (kg) should be utilized. • The law expects that each honey segment put in a holder (segment comb or cut comb honey) be set apart with the net base load in ounces and the name of the maker or dealer. Net weight is normally 1 ounce, not exactly the gross weight (1 ounce being taken into consideration the heaviness of the part's wood.

There are three essential ways to deal with promoting the honey harvest. With a couple of colonies, you can sell all of your honey from the home to family members, neighbors, and individuals from the local area. As your activity expands, you should track down other market outlets. Neighborhood food, organic product stands, well-being food stores, and side-of-the-road markets are expected outlets to investigate. When considerably bigger amounts are created, the beekeeper might need to pack in enormous mass containers and sell straightforwardly to discount vendors and packers. This technique for selling is the most un-beneficial. Many bigger honey makers have a place with the Sioux Honey Helpful.

Most little beekeepers genuinely offer their clients a quality pack in a perfect, alluring compartment. At the point when they work hard, their clients return routinely, and the beekeepers have no issue selling their whole yield. To have rehash benefactor age, beekeepers should sell their yield step by step throughout the year and attempt to showcase every one of the honey they produce not long after it is collected. Honey sold straightforwardly to the purchaser by the maker can return many benefits. Tragically, little beekeepers, much of the time, neglect to sell the honey harvest at a fair cost. Many neglect to consider the requests on their significant investment and, by and large, venture since they consider their activity a pleasant side interest instead of a business. If you sell your honey at too low a value, you harm yourself and different beekeepers in your district. The ideal way to advance honey deals is first to create a top-notch pack, then find ways of advancing your item. A "Honey available to be purchased" sign can be shown before your home. You want to screen honey showcases in shops and stands so the presentation is spotless and alluring. Any honey that has begun to take shape should be supplanted with newly pressed honey. Honey that takes shape quickly is best made into creamed honey.

Moreover, it pays to converse with individuals about your bees and their intriguing public activity. Addressing nearby help associations and nursery clubs, as well as having articles in neighborhood papers, will improve your permeability for the end goal of advertising. Going to occasions where you can show a perception hive that stands out, give taste tests of your honey, and hand out recipes adds to fruitful advancement.

Chapter 9

Bees Diseases, Parasites, and Pests and Their Control

The principal line of the guard in shielding your states from diseases, parasites, and different ailments lies in your capacity to distinguish and perceive early side effects of these issues. The inability to perceive issues early can prompt diminished efficiency and powerless and, surprisingly, dead states.

Brood Diseases

American Foulbrood

American foulbrood (AFB) is an irresistible brood disease brought about by the spore-shaping bacterium Paenibacillus hatchlings. It is the broadest and most damaging of brood diseases. AFB does not impact grown-up bees. Paenibacillus hatchlings happen in two structures: vegetative (pole-formed bacterial cells) and spores. Just the spore stage is irresistible to honeybees. Hatchlings under 53 hours old become tainted by gulping spores present in their food. More established hatchlings are not defenseless. The spores grow into the vegetative stage shortly after entering the larval stomach and continue duplicating until the larval demise. New spores structure after the hatchling passes on. Demise regularly happens after the phone is covered, during the most recent 2 days of the larval stage or the initial 2 days of the pupal stage. Brood brushes in a contaminated settlement have dispersed and sporadic examples of covered and uncapped cells. Contaminated cells are stained, depressed, and have penetrated cappings . This "pepperbox" appearance stands out from the yellowish-brown, curved, and congruity of fixed cells of a solid brood brush. Dead hatchlings change continuously from a solid silvery white to light brown and afterward to a dull espresso brown. With American foulbrood, this variety change is uniform over the whole body. In a month or somewhere near, these dead hatchlings dry down into fragile scales that are practically dark. Each scale contains upwards of 100 million spores. The scales lie level along the lower walls of the cells, with the back segment bending mostly up the lower part of the cell. House bees can't eliminate the scales from the cells. During the beginning phases of rot — up until around 3 weeks after death — the dead hatchlings have a paste-like consistency. The phone mass might stretch out for an inch or more when a toothpick is embedded and removed, known as the "ropy stage"). At the point when passing doesn't happen until the pupal stage, pupae go through similar changes in variety and consistency as hatchlings. Furthermore, a pupal tongue stands up from the remaining parts toward the top mass of the cell; this is one of the most trademark side effects of American foulbrood, however may not always be available. A couple of dead hatchlings or pupae will be seen when the disease first contaminates the colony. Once settled, however, AFB spreads quickly through the brood region. AFB can be sent rapidly to other solid states in a similar area and even to local apiaries whenever unrestrained. Nurture bees inside the hive unintentionally feed Honey sullied with spores to youthful hatchlings, propagating the disease. As the number of brood cells increments with the sizes of dead hatchlings, which are spore repositories, housecleaning bees additionally help in spore dispersal. Honey supplies inside the brood chamber become tainted as Honey is put away in these spore-loaded cells. Bees move Honey from the brood chamber to the supers above, spreading disease throughout the hive. As the disease debilitates a settlement, the state can never again guard itself against burglars from solid colonys nearby. Spore-sullied honeys spread rapidly from one hive to another. American foulbrood is likewise communicated through the trade of brushes between hives. At the point when this disease isn't perceived in an apiary, brushes from a diseased hive unintentionally might be: (1) utilized in making parts or increments, (2) utilized in trading brood and food among hives, and (3) blended in with brushes from different hives during honey extraction. Moreover, the beekeeper's hive device and gloves might spread AFB from one hive to another. American foulbrood spores are exceptionally impervious to drying up, intensity, and compound sanitizers. These spores can stay harmful for over 40 years in brushes and Honey Along these lines, Honey should not be bought from different sources to care for bees. Possibly feed brushes of Honey, assuming that you are certain beyond a shadow of a doubt they are a sans disease. An unpracticed beekeeper shouldn't buy bees or equipment that a poor person has been inspected by an assessor or someone acquainted with the disease. Indeed, even a wanderer swarm from a tainted settlement might convey AFB.

American foulbrood can't be sent to people, meaningfully affecting Honey for human utilization. Given the diseases exceptionally infectious and crushing activity, each beekeeper ought to know the side effects and have the option to perceive AFB in its beginning phases. Assuming you suspect disease and need assistance in conclusion, contact your state apiary examination administration. To send tests for the conclusion, select an example of brood to brush around 4 inches square with an enormous number of suspect cells. The segment of brush can be enveloped by some breathable material, for example, tissue or paper towel and sent in serious areas of strength for a container. Try not to utilize aluminum foil or plastic packs. Tests squashed, wet from buildup, or rotten in light of ill-advised bundling make conclusions unimaginable. Another option is to smear the items in the suspect cells in aluminum foil. Settlements tainted with American foulbrood ought to be eradicated by consumption. Consuming the bees and the equipment is the main sure method for being liberated from this disease. The base board, hive bodies, supers, and inward and external covers can be cleaned and reused. In any case, there is no assurance that the equipment can be disinfected, and the disease might return. Before consuming, diseased states ought to be killed at night after all rummaging exercises have stopped. This should be possible by dousing the bees in the colony with foamy water. Consuming bees and equipment viewed as tainted with anti-microbial safe AFB is energetically suggested and, surprisingly, compulsory in a few Mid-Atlantic States. To consume diseased equipment, dig a pit 18 inches down and wide enough to hold all brushes and equipment to be scorched. Fabricate a fire in the pit. Set your unopened hive near the pit and drop all brushes and dead bees into the fire. After everything has been scorched and the region cleaned of little bits of brush or bees, cover the remains with soil. The saved equipment (base sheets, hive bodies, and covers) should be scratched to eliminate all propolis and wax, then scoured with a solid brush and hot, foamy water. Discard the wash water and copy the scrapings so they are unavailable to the bees. After scratching and cleaning, all prepare ment should be either fire singed or drenched in a bubbling lye arrangement. Set up your lye arrangement (sodium hydroxide) by blending 1 pound of lye in with 10 gallons of water. Heat the equipment for 20 minutes; longer openness can harm wooden parts. More vulnerable arrangements may not eliminate the entirety of the wax and propolis from the equipment. Remember that lye arrangements are harsh and can cause serious consumption. Before utilizing, read the name cautiously and see all safety measures. A blowtorch is reasonable for burning little amounts of equipment. Consume the surface until it is light brown, making a point to incorporate the corners. For huge amounts of hive bodies, brush within surfaces with lamp fuel. Stack the hive bodies with the metal rabbets looking lower on top of one another, five to eight supers high, and afterward touch off the stacks, permitting them to consume to the point of gently scorching the wood. Another methodology is to fill the stack with rolled sheets of paper sprinkled with lamp fuel. When you are done, put an external cover on top of the stack to cover the fire. Anti-infection agents like Terramycin® (oxytetracycline HCL) have been utilized as a preventive measure and a treatment against American foulbrood. These anti-microbial don't kill Paenibaccillus hatchlings spores however forestall or postpone their development when present in low fixations in the food taken care of by workers to defenseless hatchlings. While this treatment permits individual hatchlings to make due, it fails to address the harmful spores in the debased equipment. In this manner, the disease ordinarily returns once the drug-taking care stops. Likewise, expanding quantities of colonys and entire beekeeping activities are being viewed as contaminated with AFB impervious to the anti-toxin Terramycin. New anti-infection agents are being explored for treating foulbrood. Notwithstanding, definitely, after some time, these too will have diminished viability because of the intrinsic capacity of diseases to foster protection from drugs.

European foulbrood (EFB), one more bacterial brood disease, is brought about by Melissococcus pluton. This disease is viewed as a pressure disease and is most common in spring and late spring — the period while brood raising is at its level, and the earliest brood is seldom impacted. European foulbrood often starts to vanish with a nectar stream and may vanish completely for the equilibrium of the year, or it might return during nectar deficiencies in the late spring or fall. Once in a while, the disease stays dynamic all through the whole scavenging season. All positions of bees are vulnerable; business strains contrast in powerlessness. The disease and its side effects are likely exceptionally capable because a few different sorts of bacteria are often present in dead and passing on hatchlings. EFB typically doesn't kill the settlement. However, a weighty infection will genuinely diminish populace improvement. European foulbrood mostly kills hatchlings 2 to 4 days old while curling in the cells' lower part. Dissimilar to American foulbrood, most hatchlings bite the dust before their cells are covered. A patchy example of covered and uncapped cells grows just when EFB arrives at serious extents. Periodically, pupae pass on from the disease. The main side effect of EFB is the variety change of the hatchlings. They change from an ordinary magnificent white to yellow, then, at that point, brown, and lastly, grayish dark. Hatchlings likewise lose their stout appearance and look undernourished. In such cases, the remaining larval parts seem wound or snaked in the lower part of the cell. They structure a flimsy brown or blackish-brown scale, showing unmistakable lines where their breathing cylinders are found. As of late dead hatchlings are seldom ropy. Scales can be removed effectively from the cells since they don't stick firmly to the cell as in American foulbrood. Bees eliminate the scales by upgrading or taking care of a nectar stream.

Melissococcus pluton doesn't frame spores. However, the organic entity frequently overwinters on brushes. It acquires a section into the hatchling in debased brood food and duplicates quickly inside the stomach of the hatchling. Not all tainted hatchlings bite the dust from the disease. Some grow typically and void the microbe or spew the microorganisms onto the underside of the cappings, which then become wellsprings of the disease. Sometimes, European foulbrood can be wiped out by requeening settlements with a youthful queen. Requeening gets two things done: it gives the settlement a more productive queen and allows a delay between brood cycles that permits the house bees to eliminate diseased hatchlings from their cells. Anti-infection agents, such as Terramycin, have been used to treat this disease.

Chalkbrood

It is a parasitic brood disease of honeybees brought about by the spore-shaping growth of Ascosphaera apis. Specialists, drones, and queen hatchlings are defenseless. Dead hatchlings are pale white and typically covered with fibers (mycelia) that have a cushioned, cotton-like appearance. These embalmed hatchlings might be mottled with brown or dark spots, particularly on the ventral sides, because of spore growths or fruiting collections of the parasite. However, diseased hatchlings can be found throughout the brood-raising season, which are most pervasive in pre-summer when the brood home is quickly extending. Chalkbrood generally vanishes or declines as the air temperature climbs in the late spring. Impacted hatchlings are found on the brood home's external edges, where there are hardly any medical attendant bees to keep up with the brood home temperature. Brood cells can be either fixed or unlocked. Youthful pupae or, as of late, fixed hatchlings are generally vulnerable. Contaminated hatchlings, loosened up in their cells, are taken out by nurture bees 2 to 3 days after side effects initially show up. Dead hatchlings (mummies) are often tracked down before the hive, on the base load up, or in the dust trap. In solid settlements, most of these mummies will be disposed of by working drones beyond the hive, consequently lessening the chance of reinfection from those who have passed on from the disease. Spores of the organism are ingested with the larval food. The spores develop in the rear stomach of the honeybee hatchling, yet mycelial (vegetative) development is captured until the hatchling is fixed in its cell. At this stage, the hatchling is around 6 or 7 days old. The mycelial components get through the stomach wall and attack the larval tissues until the whole hatchling is survived; this interaction mostly takes 2 to 3 days. Spores stay destructive for quite a long time.

Consequently, tainted bits of equipment, particularly brood brushes, is a supply for additional contamination. Chalkbrood, for the most part, doesn't obliterate a settlement. At the point when the disease is serious, in any case, it can forestall typical populace development and excess honey creation. Research has shown that the spores are effectively passed from one honeybee to another. Hence, floating and burglarizing bees are possible vectors of the disease. The two workers and queens from contaminated states can transmit disease to sound settlements. Settlements that take care of dust gathered from tainted colonys will likewise get the disease.

Nosema disease, a dark executioner, is brought about by a spore-framing protozoan (Nosema apis) that attacks the gastrointestinal systems of honeybee workers, queens, and drones. Spores of the disease are ingested with food or water by the grown-up honeybee. The spores grow and increase inside the covering of the midgut. Many spores are shed into the gastrointestinal system and killed in excrement. Harm to the gastrointestinal system might deliver diarrhea and debilitate the bees. Subsequently, the useful existence of the laborer is abbreviated, and its capacity to create brood food diminishes, hindering brood creation and colony advancement. At the point when queens become tainted, egg creation and life expectancy are decreased, prompting supersedure. Tainted workers, compared to sound workers, may poo in the hive. Diseased colonys generally have expanded winter misfortunes and diminished honey creation. The deficiency of queens in colonys recently began from bundle bees is the most serious impact of the disease. Even though nosema is a typical disease of bees, it, by and large, slips through the cracks in the apiary since it doesn't create signs or side effects that are effortlessly perceived under field conditions. The presence of the disease is normally acknowledged once the vast majority of the bees in the settlement are contaminated. The main positive approach to recognizing the disease is analyzing grown-up bees. The diseased bees' rear stomach and intestinal system are powdery white or smooth white. Solid bees, then again, have golden or clear intestinal systems. Likewise, individual roundabout choking influences of a sound honeybee's stomach is apparent, though the stomach of a tainted honeybee might be enlarged, and the tightening influences are not apparent. The minute assessment of macerated stomach tissue for the presence of spores best identifies the disease. Nosema disease is most common in the spring.

The seriousness of the disease differs among colonys, geo-realistic locales, and from one year to another. In overwintered states, restricted tainted bees might poop on the brushes and contaminate sound bees as they clean them in the spring. Food stores and dirtied delivering confines are different wellsprings of contamination. Tainted bundle bees, parts, and debased equipment spread spores. Brushes from debilitated states that passed on throughout the colder time of year frequently have nosema-polluted dung. Establishing bundles or divisions on this equipment in the spring hampers settlement improvement and frequently brings about queenless ness. Queens might become tainted from different sources after they rise from the queen cells or are delivered in mating cores.

When the disease is extreme, state populaces might become exhausted and, in the end, lessen to a modest bunch of bees and a queen. This is frequently alluded to as "spring waning." Brood raising permits the state to recuperate in colonys that are just somewhat impacted. Colony constrained during winter or harsh climate in the spring energizes nosema disease development. Winter purging flights empower bees to crap outside, forestalling spore contamination inside the bunch. Nosema-wiped-out bees frequently fly from the hive at minimal flight temperatures, most likely because they are under pressure. Since they are feeble, they drop to the ground, become chilled, and can't return to the colony. Wiped-out bees are, in this way, disposed of from the state. The force of contamination ordinarily dies down in April as field flights start and brood development speeds up. Brood development, the settlement's essential normal guard against nosema, replaces contaminated bees with youthful sound bees. Assuming nosema is as of now present, any break in the brood-raising cycle will probably expand the occurrence of the disease, particularly in late winter. Recently hived bundle bees are truly vulnerable to nosema. During the initial 3 weeks following establishment, when the state has no arising youthful bees, the disease can spread quickly through the old grown-up populace and to the queen. The best protection against nosema is to major winter areas of strength for just a lot of Honey in the legitimate position and with youthful, energetic queens. Many synthetic compounds have been tried to control the disease, yet Fumidil-B (fuma Gillin) has demonstrated power. Fumagillin is particularly powerful in stifling nosema in overwintered settlements and recently settled bundles. Nonetheless, since fumagillin doesn't influence spores of the nosema parasite, treatment with this medication won't remove the disease from the settlement.

Parasitic Mites

Varroa Mite

The varroa parasite, Varroa destructor, is thought of by quite a few people as the most serious illness of honeybees. It currently happens almost around the world. This outside parasite benefits from the hemolymph (blood) of grown-up bees, hatchlings, and pupae. Weighty parasitism brings about weighty honeybee mortality and ensuing debilitating of the state and can prompt colony demise. This is particularly evident if certain infections are available and communicated by varroa parasites. This vermin is an outer parasite apparent to the unaided eye. The female vermin is brown to rosy brown in variety and is about the size of a pinhead. Guys are somewhat more modest and light tan in variety. Grown-up guys don't take care of and are not tracked down beyond brood cells. Grown-up female vermin can live external the brood cells and are tracked down by grown-up and working drones. This conduct permits them to attack new host settlements and endure the colder time of year in these states. The smoothed state of the female's body makes it simple for the vermin to clutch a honeybee and move effectively into the cells to create a honeybee brood. When on grown-up bees, female varroa is found predominantly on the highest point of the honeybee's chest where the wings append, between the head and the chest, between the chest and the midsection, or between covering sections of the mid-region. There the vermin can undoubtedly utilize their puncturing mouthparts to enter the exoskeleton of their host and get sufficiently close to the honeybee's hemolymph. These are likewise spots where bugs are more averse to being eliminated by the honeybee's prepping. When female vermin are prepared to lay eggs, they move into brood cells containing youthful hatchlings not long before the cells are covered. They go to the lower part of the brood cells and drench themselves in the leftover brood food. After the cells are covered and the hatchlings have completed the process of turning casings, the vermin begin benefiting from the hatchlings. They start laying eggs around 3 days after the cell has been covered. A prepared female bug lays a sum of four to six eggs. The grown-up female and its juvenile posterity feed and foster on the honeybee as it develops. After being brought forth from the egg, the juvenile parasite goes through a few formative stages, including the eight-legged protonymph and deutonymph stages. The period from egg to grown-up takes 6 to 7 days for the female and 5 to 6 days for the male. Mating happens in the brood cells before the new grown-up females arise. The grown-up guys kick the bucket after intercourse since their mouthparts (chelicerae) are altered for sperm movement as opposed to taking care of. The old female and the recently treated female posterity stay in the brood cell until the youthful honeybee arises. The grown-up honeybee fills in as a middle host and a method for transport for these female parasites. More bug posterity can develop during the drones' extended advancement time (covered time). Notwithstanding, laborer brood likewise is gone after. Queen brood is gone after just in instances of weighty pervasion. Female bugs delivered in the late spring live 2 to 90 days, reduced in the fall live 5 to 8 months. The bugs can endure something like 5 days without bees or covered brood. We don't know unequivocally how varroa parasites spread so quickly. We, in all actuality, do realize that these bugs can be spread by the development of honeybee settlements (transient beekeeping), the shipment of queens and bundle bees, and the development of colonys for fertilization rentals. Beekeepers presumably spread an invasion starting with one state and then onto the next through ordinary apiary controls. Pervasions likewise are spread because of floating (particularly floating drones) and amassing bees. Individual-creating bees, whenever plagued with one to two grown-up bugs (and posterity), may arise without apparent harm and are typically ordinary by all accounts. They may, notwithstanding, experience the ill effects of malnutrition, blood misfortune, or disease. People vigorously pervaded with in excess of five grown-up parasites (which produce upwards of 20 sprites) typically become noticeably disabled or kick the bucket in their phones without emerging. The amount of this harm because of the taking care of parasites versus the presentation of infections isn't known. At the point when grown-up bees are plagued with at least two vermin, they become anxious and fly with trouble, their life expectancy is, for the most part, more limited contrasted with unparasitized bees, and they perform assignments inadequately. On a settlement level, the side effects of a varroa bug pervasion rely upon the level of invasion and the presence of certain infections. Low-level varroa invasions are challenging to identify. Medium-to-significant level pervasions might bring about an inconsistent brood design and the presence of twisted laborer and drone grown-ups with distorted wings (which might be related to disfigured wing infection) and little mid-regions. Such bees are frequently unfit to fly and should be visibly creeping. Bees will uncap and toss out invaded brood, which can sometimes be found at the hive entrance. Parasitized pupae will seem to have little, pale or dim rosy brown spots (mature and perhaps youthful vermin) on their typically white bodies. States can become seriously weakened as parasite populaces arrive at significant levels toward the finish of the brood-raising season, particularly if certain infections are available. To date, these bugs have been largely controlled through synthetic means. Plastic strips impregnated with a compound pesticide have been utilized throughout the U.S. These strips convey a contact pesticide, implying that the vermin should interact with the strips for the material to be compelling. Be that as it may, bugs have rapidly evolved protection from these synthetics, making them pointless in a brief timeframe.

Pests

Little Hive Insect

(Aethina tumida) Our freshest honeybee bug was first recognized in Florida in the spring of 1998. Before its disclosure in the U.S., the scarab was known to exist just in tropical or subtropical areas of Africa. How it found its direction to North America is still being determined. Since grown-ups will benefit from leafy foods, particularly enamored with melon, the creepy crawlies might have been inadvertently brought into this country through a shipment of natural products from Africa. While the little hive creepy crawly isn't viewed as a serious bug in South Africa, some Florida beekeepers encountering weighty pervasions have seen the speedy breakdown of solid states. The creepy crawly is generally tracked down in Florida, Georgia, and Carolina's apiaries. They were likewise found in supers of Honey sent north from Florida and in bundles disseminated in a few states. Regions where it has effectively secured itself seem, by all accounts, to be predominantly confined to districts along the East Bank of the U.S., where sandy soil conditions permit the creepy crawly to effectively finish its life cycle. The grown-up scarab is little (around 1/3 the size of a honeybee), rosy brown or dark in variety, and covered with fine hair.

The hatchlings are little, cream-hued, and comparable in appearance to youthful wax moth hatchlings. By looking at their legs, you can separate the scarab hatchlings from wax moth hatchlings. Scarab hatchlings have three arrangements of legs simply behind the head. Wax moth hatchlings, similar to all moth and margarine fly hatchlings, have three arrangements of legs behind the head, yet what's more, they have a progression of matched prolegs that run the length of the body. Prolegs are missing in creepy crawly hatchlings. Grown-up females lay their enormous egg masses on or close to beeswax brushes. The eggs hatch in a couple of days, creating many little hatchlings. The hatchlings consume dust and wax yet will additionally eat honeybee eggs and hatchlings. They complete their larval stage in 10 to 16 days and drop to the ground, where they pupate in the dirt.

Grown-ups rise out of the dirt in roughly 3 a month. The females are fit for laying eggs roughly a multi-week after rising out of the dirt. They are great flyers and effectively spread to new settlements where they store eggs to start another age. Close perception of bug-swarmed settlements in Georgia has shown that the scarabs shut down propagation during winter. While the scarab is viewed as a minor bug in South Africa, the U.S. experience to date recommends that it be a more serious vermin in certain regions of the country. A few beekeepers have detailed that scarabs can assume control over areas of strength for even.

Notwithstanding, in most cases, it is the feeble as well as diseased settlements that are plagued and capitulate to this irritation. Likewise helpless against assault are full honey supers put away in the honeyhouse or on hives above. Honeybee escapes for extensive periods. When bugs get traction, even a couple of grown-ups can create masses of hatchlings.

As well as consuming the colony's assets, the grown-up scarabs poop in the Honey, making it mature and run out of combs. When little hive bug pervasions become weighty, queens will quit laying eggs, and the colony will diminish or flee. The entire spring and ensuing hive examinations should be finished with an eye open for this irritation. While opening a hive containing creepy crawlies, they should be visible stumbling into the combs to track down concealing spots. Grown-ups may likewise be recognized under top covers or on base sheets. Assuming an invasion is weighty, the two grown-ups and masses of hatchlings might be seen on the combs and baseboard. These hatchlings don't deliver smooth passages, webbing, or covers in the hive (as wax moth hatchlings do). Varroa vermin tacky sheets are ineffectual for use in recognizing grown-up creepy crawlies. The insects move effectively across the tacky material regardless of whether the sheets are covered with a stickier material like Tanglefoot. In any case, ridged cardboard with the paper eliminated from one side, put on the baseboard at the back of the hive, has effectively distinguished grown-up creepy crawlies. The creepy crawlies seem to look for cover in the creases. Matured Honey radiating from full supers away, ready to be separated, or on dynamic colonies indicates that hive insects might be available. The matured Honey might radiate a "rotting orange" smell. On the off chance that you track down proof of, or are worried about, the chance of, a hive creepy crawly invasion, you are encouraged to quickly contact your state apiary reviewer (Branch of Farming; see supplement). To lessen the danger of this vermin in your apiary(ies), you ought to play it safe: • keep up with areas of strength for states • keep apiaries clean of all equipment not being used • extricate Honey when it is taken out from settlements • obliterate bugs when they are recognized.

Skunks, Opossums, and Raccoons

In certain territories, skunks, opossums, or raccoons can be a serious irritation and, surprisingly, a danger to effective beekeeping. Being insectivorous (bug-eating), these hunters will strike the honeybee yards daily, consuming enormous quantities of bees and hampering the advancement of solid settlements. While such goes after are most normal in the spring or fall, they additionally can happen all through the mid-year. To catch their prey, skunks scratch at the hive entrance; when the workers emerge to explore the aggravation, they are wrecked and eaten. Skunks bite on the bees until every one of the juices is polished off, then let out the remaining parts. A fruitful skunk will rehash the cycle a few times and may take care of it at the hive entrance for an hour or more. As well as quickly exhausting the grown-up honeybee populace, these hunters make a state exceptionally protective since they generally return many evenings. Besides the front of the hive being scratched up and sloppy, the grass before the hive will be pressed down or destroyed, and little heaps of bitten-up or pooed honeybee parts will be noticeable. Solid states occasionally put forth a valiant effort, yet more vulnerable colonys ordinarily succumb. Like this, keeping up areas of strength for with is an incomplete dissuade lease to hunter assault. These creatures likewise might be deterred by screens or queen excluders joined to the front of the hive and covering the entry or by raising the hive. These gadgets hamper scratching at the front entry, and on the off chance that a hunter scales the screen over the entry, its midsection becomes helpless against stings. Fencing the honeybee yard or putting the settlements on stands would be a compelling procedure, yet the expense might make it prohibitive. Moving your bees to another area is another methodology thought about as unfeasible often. At present, no synthetic anti-agents or poisons are marked for controlling hunters. Since these are classed as fur-bearing creatures, they are safeguarded besides during the yearly catching season (late harvest time). Nonetheless, the landowner can kill wild creatures that participated in the material destruction of developed crops, natural product trees, vegetables, animals, poultry, or beehives.

Summing Up

Congratulations on making it to the end of this book! We hope that you have found the information contained within these pages to be both informative and practical. Beekeeping is a fascinating and rewarding hobby, and we hope that we have provided you with a solid foundation for beginning your own beekeeping journey. Throughout this book, we have covered a wide range of topics related to beekeeping, from the basics of hive construction and management, to the intricacies of honey production and disease prevention. We have also delved into the fascinating social structure of honey bees and explored their vital role in pollinating crops and maintaining the health of our planet's ecosystems. It is our sincere hope that this book has inspired you to take up beekeeping, or to deepen your existing knowledge and skills. Remember, successful beekeeping requires dedication, patience, and a deep respect for the bees and the natural environment in which they thrive. As you embark on your beekeeping journey, always prioritize the health and wellbeing of your bees. Regular inspections, disease prevention measures, and proper hive management are essential to keeping your bees healthy and productive. It is also important to stay up-to-date on the latest research and best practices in beekeeping, and to seek guidance from experienced beekeepers if you encounter any challenges or concerns.

Finally, we want to thank you for joining us on this exciting adventure into the world of beekeeping. We hope that this book has been a valuable resource for you, and we wish you all the best in your future beekeeping endeavors. Happy beekeeping!

I hope you enjoyed reading my book! I would love to hear your thoughts and feedback. Your input is invaluable to me as we work to improve and create better content for you!

If you liked this book, please let us know your feedback on the platform where you purchased this book. Your feedback will not only help me to improve, but it will also help other readers to decide if my book is right for them.

Thank you for your time and support.

BEEKEEPING 101
FOR BEGINNERS

Take your Beekeeping to the next level